BIOMES AND THEIR IMPORTANCE

BIOMES AND THEIR IMPORTANCE

Edited by

Dr. Esha Rami

Delve
PUBLISHING

www.delvepublishing.com

Biomes and Their Importance

Dr. Esha Rami

Delve Publishing
4164 Lakeshore Road
Burlington, ON L7L 1A4
Canada
www.delvepublishing.com
Email: orders@arclereducation.com

ISBN: 978-1-77956-355-2 (Hardcover)

ABOUT THE EDITOR

Dr. Esha Rami is presently working as an Assistant Professor, in the Department of Life science, Parul Institute of Applied Science, Parul University, India. She Did her Post graduated from Ganpat University, Ph.D. in biotechnology from Hemchandracharya North Gujarat University, in the year 2015. She has authored a number of national and international publications in reputed journals.

TABLE OF CONTENTS

LIST OF FIGURES

ABSTRACT

This book introduces the concept of biomes, expansive ecological regions defined by climate and dominant flora, offering insights into their distribution and ecological roles. Covering tropical forests, grasslands, deserts, tundra, and wetlands, the book explores the distinct characteristics and functional traits of organisms within each biome. Although biomes span varied genetic compositions, the functional traits of dominant species within each biome type show remarkable similarities across regions. The text emphasizes that wetland biomes, such as marshes and shallow-water wetlands, require active management to prevent succession to forest wetlands, highlighting the importance of biodiversity conservation through dynamic, rather than passive, preservation strategies. This resource is designed for students and researchers interested in understanding the distribution and ecological significance of biomes and the necessity of sustainable management practices to protect biodiversity.

PREFACE

Large regions of our terrain known as biomes are distinguished by a specific kind of ecology, which is frequently expressed in terms of the flora. The climate of an area determines its biome and deserts can be found in very dry places, grasslands can be found in slightly wetter locations, and forests predominate in wetter areas. Despite frequently having highly varied genetic compositions, the functional traits of dominant organisms in a given biome type are similar throughout very different regions. The biomes covered in this book are the tropical forests of the equatorial regions, the grasslands of the prairies and the tropical savanna, the deserts of the arid locations, and the tundra habitats of the polar regions and high mountains. The wetlands, with the environments of rivers and lakes, are dispersed over the continent rather than lying in specific climatic zones. Certain types of wetlands such as marshes and shallow-water wetlands can be evolved in specific areas. They require active and sustainable management to stop the succession to forest wetlands. A mosaic of several stages growing in succession can be used to preserve a variety of wetland ecosystems. The maximum amount of species variety is therefore achieved. Therefore, conservation goes beyond mere preservation. We need to actively participate in their management to make sure that their enormous biological wealth is preserved. However, for this kind of management to be effective, wetlands and other biomes must be accurately understood, and this can only be done by thoroughly studying these unique ecosystems.

This book provides and introduction to various biomes that can be useful to all studies who are interested in knowing the distribution of biomes and their importance.

INTRODUCTION

Wetlands and other biomes can be impacted by climate change and global warming, particularly the bogs in the north where temperatures are predicted to rise more quickly. Bogs will get dried up, peat will cleave, and there will likely be an additional carbon input into the atmosphere, which can worsen the climate change problem. Coastal wetlands and others may be impacted by rising sea levels. Lakes are classified as heterotrophic if the energy available to their food webs comes from organic material brought to the lake from terrestrial sources, and autotrophic if their food webs are involved in plant production or growth. The excreta of terrestrial animals, leaves, pine needles, fruits, seeds, pollen, and dissolved organic compounds from soil-based organic material are examples of terrestrial organic materials. The carbon sequestered by photosynthesis in autotrophic lakes surpasses the carbon utilized for respiration by the biotic communities. The decomposers in these lakes provide the food energy that passes through the food web. More energy is expended in respiration in the heterotrophic lake than is generated by photosynthesis or autotrophic. Preserving the tundra will require enforcing precautions on a global scale. Certain species might need to be managed to prevent the overuse of resources, and specific environments would need to be changed to guarantee the survival of some sensitive species. The future of the tundra is in our hands, despite the fact that it is valuable to humanity. Breeders can preserve a high level of genetic diversity by selecting a variety of forms within a species.

This book provides us with some information about the major biomes that include varied environmental conditions. Some attempts were made to explain complicated concepts in simple language keeping in view the readers at all levels.

Chapter **1**

INTRODUCTION TO BIOMES

Contents

1.1 INTRODUCTION

The seabed that borders a country's coast is referred to as its continental shelf under the United Nations Convention on the Law of the Sea (UNCLOS) (Rajan, 2018). 'Territorial waters' is another name for the waterways(Reeves, 1930). Coastal countries have authority and control over the waters above the continental shelf, which are also referred to as territorial waters. The geological and legal definitions of a continental shelf under UNCLOS are very different. Volcanic islands with human habitation, like the Azores and the Canary Islands, have a legal continental shelf even though they don't have one geographically(Kueppers & Beier, 2018). A portion of coastal seas are pelagic seas, which are found beside open waters. The South China Sea (Figure 1.1) and the North Sea are two examples of pelagic seas(Pauly & Liang, 2020).

Figure 1.1 South China Sea

Source: https://commons.wikimedia.org/wiki/File:South_China_Sea_claims_map.svg,*Creative Commons License.*

Coastal waters can be classified as semi-enclosed or enclosed. The Hudson Bay, the Gulf of Mexico, and the Gulf of Carpentaria are semi-closed seas, whereas the Black, Baltic, and Mediterranean Seas are examples of enclosed seas. Certain oceans can change over time, such as the Wadden Sea, a pelagic sea that was eventually contained by human activity. Certain coastal seas may be deemed estuaries due to their exposure to sediment inputs and river flows.

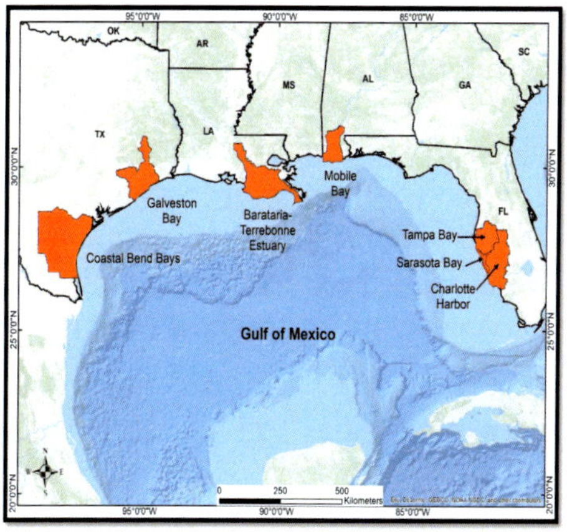

Figure 1.2 Gulf of Mexico

Source: McCarthy, M.J.; Otis, D.B.; Méndez-Lázaro, P.; Muller-Karger, F.E. Water Quality Drivers in 11 Gulf of Mexico Estuaries. Remote Sens. 2018, 10, 255. https://doi.org/10.3390/rs10020255,*Creative Commons License.*

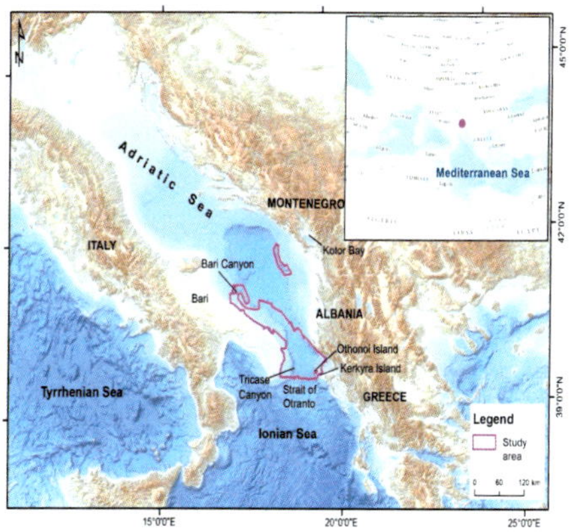

Figure 1.3 Adriatic Sea

Source: Prampolini M, Angeletti L, Castellan G, Grande V, Le Bas T, Taviani M, Foglini F. Benthic Habitat Map of the Southern Adriatic Sea (Mediterranean Sea) from Object-Based Image Analysis of Multi-Source Acoustic Backscatter Data. Remote Sensing. 2021; 13(15):2913. https://doi.org/10.3390/rs13152913,*Creative Commons License.*

The Gulf of Mexico (Figure 1.2), Mississippi River, Beaufort Sea, Mackenzie River, and Adriatic Sea (Figure 1.3) are a few examples of these estuarine-influenced waters. Polar seas that border the shores of the Arctic and Antarctic are a distinct kind of coastal sea due to their harsh environmental conditions and largely pristine state. For the majority of human history, coastal seas, especially estuaries and nearby coasts have been essential to human progress and are primarily impacted by human activity. Coastal waters have always been the main areas for commercial fishing. Their seafloor and depth allow them to offer shelter from both extreme weather and mild storms. In contrast to the sparser life found on the deeper abyssal plains, the coastal seas are home to an abundance of life due to sunlight spreading through the shallow waters over the continental shelf. The majority of marine commerce, including mining and hydrocarbon production, also occurs on continental shelves.

1.2 COASTAL AREAS

A growing portion of the global population resides in coastal areas that are close to coastal seas(Martínez et al., 2007)economic and social importance of the coasts at a global scale. We defined coastal regions to range from the continental shelf (to a depth of 200 m. The majority of the world's largest cities are situated close to natural harbors and have port infrastructure. On the flip side, smugglers, migrants, and military invaders all use the coast as a frontier. Warm water and coastal beaches can draw large numbers of visitors. Human-induced impacts on coastal seas provide numerous environmental issues due to their growing populations and uses. Records from all coastal seas, for instance, show that marine species have declined, water quality has drastically dropped, and seagrass habitats have been decimated. Coastal waters are facing an uncertain future due to a confluence of factors including overexploitation, nutrient pollution, and climate change. Petrochemicals and biological waste are examples of land-based and marine pollution, and climate change raises sea levels, endangering coastal ecosystems. Complex ocean ecosystems are changing due to a combination of factors including habitat degradation, overfishing, rising acidification, warming of the ocean, and huge fertilizer runoff(Nyström et al., 2012)marine ecosystems have shifted into contrasting states having reduced ecosystem services (hereafter called degraded. Complex marine food webs with fish and mammals are giving way to ecosystems dominated by disease, algal blooms (Figure 1.4), bacteria, and jellyfish(Pauly et al., 2009)or that outbreaks are more frequent, although much uncertainty surrounds the issue, due to the scarcity of reliable baseline data. Numerous hypotheses have been proposed for the individual increases or outbreaks that are better documented, but direct experimental or manipulative studies at the ecosystem scale cannot be used for testing them. Thus, ecological modeling provides the best alternative to understand the role of jellyfish in large fisheries-based ecosystems; indeed, it is an approach consistent with new ecosystem-based fisheries management practices. Here, we provide an overview of online databases available to ecosystem modelers and discuss general aspects and shortcomings of the coverage of jellyfish in these databases. We then provide a summary of how jellyfish have been treated and parameterized by existing ecosystem models (specifically focusing on 'Ecopath with Ecosim' as a standard modeling toolset. Understanding the current and potential future ecosystems requires an understanding of coastal seas and their original ecosystems.

Figure 1.4 Algal blooms

Source: Kislik C, Dronova I, Kelly M. UAVs in Support of Algal Bloom Research: A Review of Current Applications and Future Opportunities. Drones. 2018; 2(4):35. https://doi.org/10.3390/drones2040035,*Creative Commons License.*

Situated on the European continental shelf, the North Sea is connected to the Atlantic Ocean on both the north and south sides by the Norwegian Sea and the English Channel(Emeis et al., 2015)complex interactions within the shelf seas hinder the identification and unambiguous attribution of observed changes to drivers. These include variability in the climate system, in ocean dynamics, in biogeochemistry, and in shelf sea resource exploitation in the widest sense by societies. Observational time series are commonly too short, and resolution, integration time, and complexity of models are often insufficient to unravel natural variability from anthropogenic perturbation. The North Sea is a shelf sea of the North Atlantic and is impacted by virtually all global and regional developments. Natural variability (from interannual to multidecadal time scales. Along with being a vital shipping channel, the North Sea is home to significant fisheries. The sea produces a lot of hydrocarbons, is a popular destination for travel, and generates electricity from wind and waves. The neighboring coasts of the North Sea are heavily populated and heavily industrialized. Overfishing, dumping, dredging, and pollution from industry and agriculture are some of the environmental effects on the water. Its food web is being impacted by climate change, which is pushing its fisheries northward and allowing new species to enter from southern waters. In the North Sea, the European Union has been implementing oceanic spatial planning and coastal zone management, and there is strong regional collaboration. The Netherlands, Germany, Denmark, and the North Sea are all bordered by the Wadden Sea. It is a pelagic sea that has been transformed into a semi-enclosed sea. For many years, the Wadden Sea has been a vital transportation and food source for the nearby towns. China borders the South China Sea on the north; the Philippines borders it on the east; Malaysia, Singapore, Indonesia, and Brunei border it on the south; and Vietnam, Thailand, and Cambodia border it on the west. When the global sea level was hundreds of meters lower during the previous ice ages, the water was above a submerged continental shelf. Numerous small islands, atolls, cays, reefs, shoals, and sandbars can be found in the South China Sea; the majority of these are underwater during high tide and uninhabited. The Spratly Islands are one of these archipelagos. Adjacent nations that support various management systems, contest jurisdiction over the South China Sea.

Figure 1.5 Coral reefs

Source: Oren A, Berman O, Neri R, Tarazi E, Parnas H, Lotan O, Zoabi M, Josef N, Shashar N. Three-Dimensional-Printed Coral-like Structures as a Habitat for Reef Fish. Journal of Marine Science and Engineering. 2023; 11(4):882. https://doi.org/10.3390/jmse11040882,*Creative Commons License.*

1.3 ECOSYSTEMS

The sea is a vital ecosystem that supports coral reefs (Figure 1.5), mangrove forests (Figure 1.6), and seagrass beds(Harborne et al., 2006)seagrass beds and mangroves provide important goods and services both individually and through functional linkages. A range of anthropogenic factors are threatening the ecological and economic importance of these habitats and it is vital to understand how ecosystem processes vary across seascapes. A greater understanding of processes will facilitate further insight into the effects of disturbances and assist with assessing management options. Despite the need to study processes across whole seascapes, few spatially explicit ecosystem-scale assessments exist. We review the empirical literature to examine the role of different habitat types for a range of processes. The importance of each of 10 generic habitats to each process is defined as its "functional value" (none, low, medium or high.

Figure 1.6 Mangrove forests

Source: Purwanto AD, Wikantika K, Deliar A, Darmawan S. Decision Tree and Random Forest Classification Algorithms for Mangrove Forest Mapping in Sembilang National Park, Indonesia. Remote Sensing. 2023; 15(1):16. https://doi.org/10.3390/rs15010016,*Creative Commons License.*

This ecosystem is thought to hold one-third of all marine biodiversity on Earth. The loss of mangrove forests, overfishing, destructive fishing methods, degradation of coral reefs, destruction of seagrass beds and wetlands, and degradation of coral reefs are some of the environmental effects. For instance, development, tourism, and shrimp farms have destroyed the majority of mangrove forests. An inland sea that is brackish is the Baltic Sea. With the Gulf of Finland and the Gulf of Bothnia as its two tributaries, this sea resembles a riverbed and can be assumed as the estuary where all rivers pour into it. However, its salinity is lower than that of the Atlantic Ocean. The water flows in the Baltic Sea are complicated; lighter, saline surface waters enter the sea from the north and flow in the opposite direction, while brackish surface waters discharge into the North Sea. Because the Baltic Sea is stratified, only bacteria can thrive in the fluctuating dead zone that occurs along up to 25% of the seafloor where the heavier, saline waters go toward the bottom. The Baltic Sea (Figure 1.7) is home to both marine and freshwater species, however their diversity is quite low. For decades, algal blooms have occurred every summer. Adjacent areas' agricultural contamination has exacerbated eutrophication and prompted these blooms.

Figure 1.7 Baltic Sea

Source: Jetoo S, Kouri J. Voicing and Visualizing Change: Perceptions of Environmental Heritage in the Baltic Sea Region. Heritage. 2021; 4(3):1566-1589. https://doi.org/10.3390/heritage4030087,*Creative Commons License.*

The temperature and vertical stratification of the Baltic Sea, as well as the size and length of dead zones and algal blooms, are expected to rise due to climate change. All of the Baltic Sea's neighboring countries have been working together more closely to control the waterway, utilize its resources, and manage the sea since the fall of the Soviet Union. The Mediterranean region, which has hot, dry summers and cool, rainy winters, almost completely encloses the Mediterranean Sea, which is connected to the Atlantic Ocean. For instance, because of the small link to the Atlantic Ocean, tides are extremely restricted. Water flows are driven by evaporation, which outpaces precipitation and river flows. The eastern side has particularly high evaporation, which lowers the water level and raises salinity in that direction. Throughout human history, the sea has played a significant role in the development of both industrial and agricultural sectors.

1.4 Climate change and estuaries

The region around the Mediterranean Sea is particularly susceptible to climate change(Nicholls & Hoozemans, 1996). The changing climate regime will result in rising sea surface temperatures and salinities. Erosion, sediment deposition, drought, desertification, and flooding can all intensify or change depending on the local conditions. A significant source of revenue in the Mediterranean and South Atlantic regions is beach and coastal tourism. In order to support tourism and other industries like fishing, these regions' continued economic viability may depend on the preservation of their coastal and marine ecosystems. The plants need to be able to survive in saltwater and in generally wet, low-oxygen soil. Many marine organisms have to be able to withstand being exposed to the air, sun, wind, and possibly rain at low tide twice a day. These are difficult circumstances with frequent abrupt changes. Estuaries are regions such as bays, and inlets where freshwater from rivers and streams combines with saltwater from the ocean.

Figure 1.8 Salt marshes

Source: Zeremski T, Tomić N, Milić S, Vasin J, Schaetzl RJ, Milić D, Gavrilov MB, Živanov M, Ninkov J, Marković SB. Saline Soils: A Potentially Significant Geoheritage of the Vojvodina Region, Northern Serbia. Sustainability. 2021; 13(14):7891. https://doi.org/10.3390/su13147891,*Creative Commons License.*

Salt marshes (Figure 1.8) are located on the boundaries of estuaries and on protected shorelines(Leonardi et al., 2018). They are widespread along the Gulf of Mexico and the whole east coast of the United States, while they are less prevalent on the west coast. Salt marshes are very rare along the Pacific Coast due to the rocky shoreline of the majority of the coastline, with the exception of the two main estuary/marsh systems, Puget Sound in Washington and San Francisco Bay in California(Cordell et al., 2009). They are also found on every continent, except for Antarctica, and in other coastal areas ranging from the Arctic tundra to the tropics. Between New Jersey and northern Florida, there are vast expanses of salt marshes, especially along the coasts of Georgia and the Carolinas.

Because of the gravitational influence on the water, the lowest parts of the marshes on the east coast of the United States alternately flood and drain twice daily(Xin et al., 2022). The moon's phase, along with a host of other variables including the shoreline's form and wind direction and strength, influence the precise height and timing of the tides at any specific location. The Bay of Fundy (Nova Scotia), where the gap between high and low tides is about 50 feet, is one example. A salt marsh ecosystem is quite unpredictable because freshwater from the ocean and inflowing freshwater constantly mix in the estuary. The water in a marsh directly next to the ocean will be saltier than the water in a marsh further upstream in an estuary that gets a lot of freshwaters. The habitat and resident species are predicted by the ratio of freshwater to saltwater. For instance, a brackish marsh is found in an estuary's section where freshwater has more of a diluting effect on the saltwater. Freshwater marshes that are still impacted by the tides but have lower salinity content can be found further up the estuary. Freshwater wetlands that are not affected by the tides can be found much farther upstream. As the water in the marsh gets fresher, there is a simultaneous transition to fewer salt-tolerant plants and animals. The amount of recent rainfall and the stage of the tidal cycle both affect salinity. Anything that lives in a salt marsh needs to be able to withstand the extreme variations in environmental conditions, which can range from nearly freshwater to full seawater. Similar fluctuations occur in the amount of dissolved oxygen as water covers and uncovers the marsh. Seasonal variations in temperature are another feature. In winter, the water is warmer than the air, while in summer, the air is much warmer. Salt marshes are thriving ecosystems despite the harsh and unpredictable circumstances that plants and animals must endure(Ameixa & Sousa, 2020).

1.5 LIFE FORMS IN SALINE ENVIRONMENTS

Only found in shallow intertidal zones, certain grass species have developed some features to cope with salty soil and saltwater. These are known as halophytes. They are also referred to as hydrophytes since they must cope with spending some of their time submerged in water. Not only do plants have to withstand being submerged in water and salt, but they also face stress from the marsh soil's low oxygen content. This area of east coast marshes is covered in *Spartina alterniflora*, or cord grass(McKee & Patrick, 1988). As one ascends through the marsh to regions that are only impacted by the ocean during extremely high tides, the variety of plant species grows. There are a variety of kinds of grasses, herbs, and shrubs in the high marsh, each of which is suited to a particular zone. Salt marshes are known for their zoning, whereby distinct plant species can be found at varying altitudes based on their capacity to withstand immersion in saltwater and to contend for available space. When the tide is out, there is no food for marine animals that get their nourishment straight from the sea, thus they must find another way to stay moist during low tide. Except for a handful, most marsh animals are dormant during low tide. Uca fiddler crabs feed during low tide and stay inside their burrows during high tide to avoid being eaten by fish. Amphipods and isopods, which are smaller crustaceans, have the ability to stay active on the marsh surface during low tide. Additionally, the salt marsh mussel, or *Geukensia demissa*, often opens its shell to allow for air to get through. Almost all inhabitants of the salt marsh coordinate their life cycles and reproductive processes with the lunar phases and tides. Fiddler crabs release their young into the water, where they are carried by tidal currents, during the highest tides, which occur during the full and new moon(Salmon & Hyatt, 1983).

***Figure 1.9* Wet mudflats**

Source: Ramos-Fuertes A, Marti-Cardona B, Bladé E, Dolz J. Envisat/ASAR Images for the Calibration of Wind Drag Action in the Doñana Wetlands 2D Hydrodynamic Model. Remote Sensing. 2014; 6(1):379-406. https://doi.org/10.3390/rs6010379,*Creative Commons License.*

Wet mudflats, (Figure 1.9) also known as sand flats, are created when tides dump sediments, soil, fine-grained clay, sand, and silt from rivers and shorelines across low-lying terrain. This process leads to the progressive development of salt marshes. Salt marshes arise in flat locations that drain slowly. The sediment is stabilized by the formation of mats of microscopic algae and bacteria, which allows the elevation of the silt to rise. As the marsh habitat grows, this process quickens as the plants take root. When plants float in on the tides and seeds are transported in by the air or water, salt-tolerant grasses appear. With the help of underground horizontal stems, or rhizomes, the plants gradually take root and spread. As they do so, the sediment becomes more stable due to the establishment of thick root systems, and they also add volume. After the grasses are established, the force of the waves is diminished. The accretion of fine particles into the marsh is caused by the decomposition of plants. Marsh grasses that can withstand the salt replace freshwater plants on the land when the marsh pushes both landward and seaward in response to rising sea levels. Peat builds up beneath the marsh from subterranean plant waste, such as roots and rhizomes(Bertness, 1988). Eventually, the level of the sediment surface is raised by peat development, reducing the frequency of flooding. At that moment, some plant species can survive that are a little less tolerant of salt and flooding. This eventually results in the creation of high marsh areas that experience less flooding from the incoming tides. Both sediment deposition and marsh accretion are facilitated by the majority of the roots and rhizomes. Sea levels are increasing more quickly than sedimentation rates in many places due to the thermal expansion of water brought on by climate change and glacier melting. If marsh accretion cannot keep up with the rise in sea level, open water may eventually displace the marshes. The overall expansion of the marsh system and the formation of the plant zonation pattern in which certain species are arranged in bands at varying elevations are the outcomes of natural sedimentation. A fully formed marsh is characterized by a network of creeks that serve as the main channel connecting the estuary to the land, serving as entry and exit points. The contributions that various animal and plant species play in the general ecology of salt marshes are observed around the world.

1.6 UNIWUE ASPECTS OF AN ESTUARINE ECOSYSTEM

Coastal regions are used by about 85% of migrating birds and ducks as a place to rest, feed, and reproduce(Stott & Olson, 1973). Salt marsh plants provide a home to a variety of estuarine organisms at different phases of their lives. During high tide, young fish migrate upward and swim within the stalks of marsh grass, shielded from bigger predators that are unable to cross the shallow creek waters. Over 2/3rd of fish that are taken for commercial purposes on the east coast are thought to inhabit the marsh-estuary ecosystem. Hence, salt marshes are regarded as nurseries, where a variety of fish and shellfish species acclimate and rely on the marsh for protection and food. For example, bluefish spawn in the ocean, but their young/offspring migrate into estuaries in the spring and stay there all summer to feed on the smaller fish that inhabit the marsh/estuarine system. Certain species, such as *Fundulus heteroclitus* live their whole lives in marshes and estuaries(Fernández-Delgado, 1989). Benthic habitats on the continental shelf range from coarse gravel, stones, and boulders to bare rock. These habitats are varied and heterogeneous. Certain plant and animal species are supported by distinct substrata, each of which has organisms that are specialized to a particular set of environmental conditions. Muddy and sandy places are ideal for organisms that live on or in sediment; examples are worms, prawns, lobsters, and flatfish. Rocky environments support a variety of animals, such as octopus, algae, anemones, sponges, bryozoans, urchins, starfish, and predatory fish, as well as more complicated physical habitats. In continental seas, the water column serves as a vital habitat for a wide variety of pelagic fish and invertebrates, including squid, jellyfish, sardines, haddock, cod, and even larger animals like sharks, dolphins, and whales.

Intertidal coastlines are rated according to a gradient of exposure, with options for both protection and exposure(Tibbetts & van Proosdij, 2013). Substrates might be sandy, muddy, stony, or rocky. The influences of weather and proximity to steady freshwater inflows cause salinity to vary significantly. The exposure to sun radiation, wind, and terrestrial organism predation varies significantly in intensity depending on the season based on the tides. Barrier reefs, mangroves, wetlands, marshes, and tidal mudflats are examples. The geology of the benthos shifts from continental crust to oceanic crust as it moves from the shelf to the bottom. Because of their frequent low sea levels, continental slopes serve as a habitat for rich mineral deposits that are carried there by marine currents, terrestrial runoff, and glacial deposition. These waters are cooler, have less light, and have higher barometric pressure. Long-lived species with slower growth and metabolic rates are typically found in the deeper slopes. Some of the ocean's richest areas for marine life are found in the comparatively shallow seas of the continental shelves. The nutrient-rich runoff from land-based rivers and streams contributes to this partially. Additionally, sunlight can reach the seafloor in shallow waters, enabling photosynthesis. Then, marine flora support invertebrates, which support fish, sharks, marine mammals, and seabirds. Cnidarians, which include urchins, starfish, and sand dollars, are among the invertebrates that are supported by marine flora.

1.7 THE CONTINENTAL SHELF

On the continental shelf, phytoplankton and zooplankton are the building blocks of life. Plankton is made up of a vast array of species, including fish larvae and eggs, foraminifera, copepods, radiolaria, and minute, floating, single-celled plants and animals. Among these, krill are well-known. Despite being the tiniest organisms on Earth, their abundance, especially in upwelling zones, makes them the primary food supply for several marine mammals and predatory fish. Certain organisms are only planktonic during the initial phases of their life cycle, whilst others are planktonic throughout their life. But plankton is not just found in the waters of the continental shelf. Seagrass meadows are found worldwide on continental shelves and are covered in thick carpets of green leaves(Asmus et al., 2022). In the marine biome, these delicate plants play a critical role as a food supply for larger species like sea turtles and dugongs as well as a shelter for extremely rare and endangered species like seahorses. Through the stabilization of sediments and the absorption of excess nutrients in eutrophic runoffs, seagrass meadows also contribute to ecological benefits.

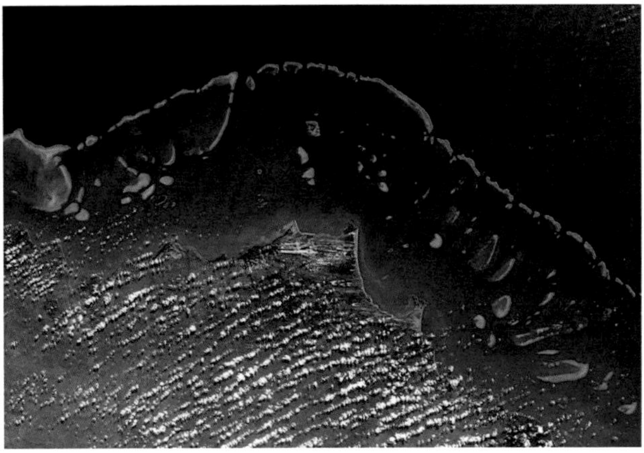

Figure 1.10 Great Barrier Reef

Source: https://commons.wikimedia.org/wiki/File:ISS-45_StoryOfWater,_Great_Barrier_Reef,_Australia.jpg,*Creative Commons License.*

Perhaps one of the less well-known benefits of coastal waters is their kelp forests. Kelp is a brown seaweed with thick herbaceous fronds on top of a strong stalk that clings to the seafloor. Although they may not be visually appealing, kelp stands are among the most significant living environments in oceans at higher latitudes. They provide food and shelter to a wide range of organisms, including seals, otters, and marine invertebrates. Coral reefs are the gem in the crown of coastal seas. These living ecosystems, which are common in the cleanest, warmest waters, are packed with diverse species. It has been hypothesized that around one-third of all known fish species worldwide reside in coral reefs. The western Pacific continental shelf is home to many coral reefs. The biggest known coral reef is found in Australia's Great Barrier Reef, which is made up of around 3,000 individual reefs rather than a single, continuous structure. In its southernmost regions, the Great Barrier Reef (Figure 1.10) extends several hundred miles offshore, while in its northernmost regions it is only a few kilometers from the shoreline.

On continental shelves, the vertebrate fauna is just as significant as the resident species(Friedland et al., 2020). There are many different kinds of marine vertebrates, including fish, sharks, turtles, porpoises, whales, dolphins, and seabirds. In the seas of the continental shelf, fish are among the most common life forms. They are important to humans as a natural resource and as both prey and predators in their respective ecosystems. Fish depend on continental shelf habitats for a variety of reasons, including spawning, feeding, and nursery grounds as well as a route for diadromous species, which complete part of their life cycle in freshwater and part in the ocean. Numerous kinds of marine mammals, including sea otters, seals, sea lions, walruses, porpoises, dolphins, and whales, can be found on continental shelves. The majority of global fisheries are situated along continental shelves. Marine fish and invertebrates have been used for trade and subsistence for many thousands of years. Fish products are now collected not just for human consumption but also for the aquarium industry, animal feed, fertilizer, and alternative medical treatments, among other secondary uses. A variety of technologies are used in fishing. In shallow, warmer waters, spear fishing and hand collection by freediving are also popular methods. Groundfish, comprising fish like cod, haddock, flounder, and sole, have long been the most frequently fished populations in U.S. coastal waters. Cod was the driving force behind North American wealth for many years; it provided food for millions and was cherished in Western culture.

For hundreds of years, it was still possible to fish these shelves, but the entrance of sophisticated ships after World War II put an end to the cod stocks. These new vessels, which were fitted with sonar, and radar could process and freeze fish on board while trawling and hauling massive nets for weeks at a time in almost any weather. To exacerbate the situation, these banks were not governed by a single country at the time since they were regarded as international waters. When it comes to natural resource extraction, the outer continental shelf (OCS) is the area of importance. The offshore portions of the US shelf that are governed by federal law are known as OCS regions(Juda, 1993). In U.S. shelf waters, four areas have been defined: the Atlantic OCS, Pacific OCS, Alaskan OCS, and Gulf of Mexico OCS. According to the statistics available from the Bureau of Ocean Energy Management, there were 23 operational extraction facilities in the Pacific OCS and 3,409 operational facilities in the Gulf of Mexico OCS in 2010. The Alaskan and Atlantic shelf sections were not exploited. After a surge of activity in the early 1980s, there has been a general declining trend in the number of established installations in recent years. OCS reserves accounted for around 30% of U.S. oil output in 2010; this percentage has been gradually rising over time. About 10% was made up of natural gas. Minerals are also quite valuable. Marine sediments have been mined off the Atlantic and Gulf coasts of the United States since the 1990s to restore beaches damaged by erosion(Arthur et al., 1999). In the United States, state governments typically regulate mining operations up to three nautical miles (5.56 nautical kilometers) from the shoreline; federal law governs activities beyond this distance, which is overseen by the Bureau of Ocean Energy Management.

1.8 PRECIPITATION

Precipitation that falls across the ground and condenses into even bigger streams and rivers is what causes rivers. One of the striking features are these ecosystems, which carry nutrients and sediment downstream, erode valleys, and eventually reach the sea or an interior lake. Numerous plant, animal, and microbial species have evolved as a result of this water movement, which also modifies the landscape and produces a range of freshwater habitats. Rivers differ greatly in size and topography which greatly affects biological traits and biodiversity. The mountain ranges, ocean currents, and patterns of global air circulation all have a significant impact on the climate, which includes temperature and precipitation patterns.

Coriolis forces greatly affect the live marine resources in coastal regions and have a major impact on river discharge patterns. Because of gravity, the denser oceanic water sinks to the bottom of the less dense surface water, resulting in the production of intermediate-density water through mixing processes. Positively buoyant materials and vertically migrating planktonic organisms collect in the slowly sinking waters.

The distribution of food particles becomes highly concentrated in this environment, and feeding organisms may significantly increase the amount of calories consumed per unit. There is a tendency for this process to proceed up the trophic levels. Coastal rivers are crucial to the dynamics of biological populations because of this. River biodiversity, and by extension the marine environment, is threatened by a number of major reasons, including agriculture, deforestation, freshwater extraction, the discharge of industrial and human waste, and the consequences of artificially altered river hydraulic characteristics. Estuaries absorb more nutrients per surface area than any other ecosystem as a result of phytoplankton blooms, which are caused by nutrient pollution. For instance, nutrient contamination in the United States has severely impaired more than 60% of coastal rivers and bays.

Large turbidity rates can also occur in many coastal waters as a result of high sediment loads, slow nutrient-rich organic matter fluxes, and excessive freshwater withdrawal from coastal rivers(Thrush et al., 2004). The structure of phytoplankton, zooplankton, and benthic communities can alter as a result of these variations in nutrients, light, and oxygen. Certain coastal ecosystems are more vulnerable to nutrient eutrophication than others due to a variety of different factors, including the dilution of nutrients by open ocean water in bays and estuaries. Coastal river flows are particularly vulnerable to environmental changes that affect ecosystems like seagrass meadows and coral reefs. The dead zone in the Gulf of Mexico, a vast region with low oxygen levels, is the coastal system impacted by eutrophication in the United States. This hypoxic area doubled in size by the summer of 1999, covering 7,722 square miles. The North, Baltic, Adriatic, and Black Seas in Europe face eutrophication problems(Rabalais et al., 2001)seasonal input of organic material and seasonally severe hypoxia/anoxia. Decreases in species richness, abundance and biomass of organisms were dramatic at the stations affected by severe hypoxia/anoxia, and lower than most literature values for similar habitats. Although there were summer/fall declines in the populations at the intermittently hypoxic site, these were not obviously related to changes in oxygen. Some macroinfauna, the polychaetes Ampharete and Magelona and the sipunculan Aspidosiphon, were capable of surviving extremely low dissolved oxygen concentrations and/or high hydrogen sulfide concentrations. Abundance of macroinfauna, primarily opportunistic polychaetes (similar to the spring. Early in the 1990s, the breakup of the Soviet Union resulted in a decline in nutrient supplies, and fertilizer use throughout Eastern Europe quickly reduced, largely reversing the eutrophication of the Black Sea. The Black Sea's nutrient eutrophication has recently increased to high levels; however, this decline was only transitory.

1.9 INTERTIDAL ZONES

Algae, suspension feeders, and predators are the main categories of organisms found in intertidal zones. Because algae serve as a key energy source and hence represent a large portion of the foundation for intertidal food webs, their proliferation has a significant impact on shore ecology. Certain shorelines may contain large masses of brown, green, and red algae, also referred to as macroalgae or macrophytes. The name microalgae refers to a group of protists that includes diatoms, blue-green algae, and other protists like euglenoids, as well as the spores of macroalgae, and it is typically present on shores where macroalgae is not abundant. Because of their extremely rapid production rates, while having a low biomass, microalgae frequently play a significant role in the food chain than macrophytes do. Brown kelps frequently sit on a rich assemblage of algae at the bottom of the littoral and sublittoral zones on the majority of exposed shorelines. The diversity of algae rises on protected coasts, where the entire shore may exhibit a thick layer of algae, which is frequently the result of brown algae growth. Red algae typically engulf the brown algae at the bottom of the coast, and on shores, the diversity of species rises toward the sublittoral. Although they are rarely seen in significant quantities, several genera of green algae can be found all over the world. The herbivorous gastropods, which include limpets, winkles, top-shells, and urchins, are specialized/adapted for the intertidal zones. Several of these organisms are found all over the planet. Herbivores consume microalgae, fungi, and algae debris. Both live and dead macroalgae may be consumed by urchins that feed sublittoral. On rocky coastlines, sessile suspension feeders like mussels and barnacles typically outnumber mobile animals. While bryozoan colonies, tunicates, sponges, hydrozoans, and polychaetes cover the bottom of rocks and macroalgae, barnacles and mussels may cover a full rock surface(George, 2012).

1.10 THE ZONAL PATTERN OF PREDATORS

In all intertidal vertical zones, predators like sea stars, and crabs are prevalent. The numerous predators that go to catch prey during high tide and then withdraw to lower zones during low tide will have different foraging ranges depending on the slope of the beach. Such predators can travel longer vertical distances in a given time on steep slopes, which may have an impact on the vertical distribution of both prey and predator. Crustaceans like crabs and gastropods like winkles typically maintain a clearly defined zonation pattern in spite of their mobility(Moschella et al., 2005)the ecological consequences of coastal defences have been less extensively investigated. Furthermore, due to their rapid proliferation, there is a growing need to understand the role of these man-made habitats in the coastal ecosystems in order to implement impact minimisation and/or mitigation measures. As part of the DELOS project, targeted studies were carried out throughout Europe to assess the ecological similarity of low-crested coastal defence structures (LCS. When the tide is high, a large variety of species are present. When the tide rises, some crabs migrate, but when it falls, they return once more. When the tide covers certain types of snails, they start migrating down-shore; however, when the tide rises again, they return to their normal area. However, as these species often stay in a small area, there must eventually be restrictions on their ability to move in the vertical plane. Crabs are one of the highly mobile species. When gastropods are moved from their usual location, they frequently find a way to return. Of the entire amount of water on the surface, freshwater lakes contain about 0.01 percent. Comparing this to the seas, which have a volume of 328.68 million cubic miles, the volume of this is about 29,989 cubic miles. Throughout the continents, there are innumerable smaller lakes and more than 920 major lakes that together comprise the freshwater lake waters.

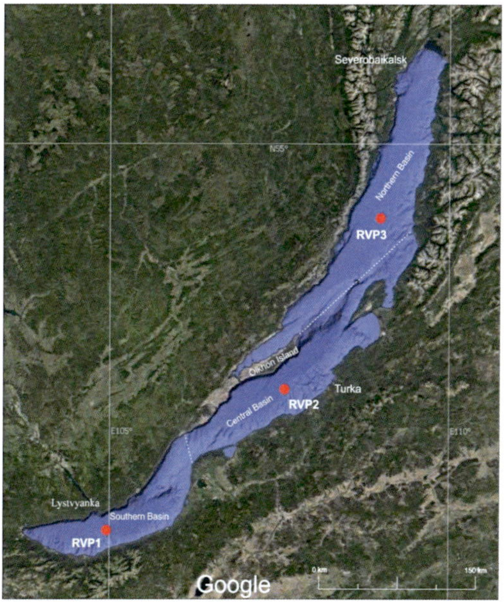

Figure 1.12 Lake Baikal

Source: Potapov S, Krasnopeev A, Tikhonova I, Podlesnaya G, Gorshkova A, Belykh O. The Viral Fraction Metatranscriptomes of Lake Baikal. Microorganisms. 2022; 10(10):1937. https://doi.org/10.3390/microorganisms10101937,*Creative Commons License.*

1.11 LAKES

At a maximum depth of more than a mile, Lake Baikal (Figure 1.12) is the deepest lake on Earth. Lake Baikal is fed by almost 300 rivers, although only the Angara River empties the lake(Arzhannikov et al., 2018). More than 2,600 recognized species of plants and animals can be found in the lake, including the deep-diving Baikal oil fish, which is the primary food source for the Baikal seal. The indigenous Buryat tribe people who reside on the eastern side of the lake depend heavily on the salmonid omul (Coregonus automnalis migratorius) that inhabits the lake. The United Nations Educational, Scientific, and Cultural Organization or UNESCO named Lake Baikal a World Heritage Site in 1996. With a length of 420 miles, Lake Tanganyika, which is surrounded by Tanzania, Zambia, the Democratic Republic of the Congo, and Burundi, is thought to be the longest freshwater lake in the world. Lake Tanganyika, with a freshwater volume of 4,534 cubic miles, is the second biggest lake in the world. At about one mile deep, it is the second-deepest lake in the world after Lake Baikal. All biological activity takes place above the depth of 656 feet since the lake's length and depth prevent it from mixing below this point and make it anoxic. Situated in the southwestern region of the African Great Rift Valley, Lake Tanganyika is a rift lake that was generated through plate tectonics. Lake Edward is one of the seven lakes in the Great Rift Valley, along with Lakes Victoria, Malawi, Turkana, Albert, and Kivu. The Ruzizi River, which enters the lake from the north, and the Malagarasi River, which enters the lake from the East, are the two main rivers that feed Lake Tanganyika. The Lukuga River on the lake's western shore is where the lake empties into the Congo River basin. There are 250 species of cichlid fish in the lake, and these fish have been the subject of studies on speciation and other aspects of evolutionary biology. Other fish species are significant as a source of food, and many cichlid fish species are popular in the aquarium business. Since the 1950s, Lake Tanganyika has been the site of commercial fishing operations. However, the fishing business has been declining, just like it has in other areas. However, since 2004, the International Union for Conservation of Nature has been monitoring the physical, chemical, and biological characteristics of the lake through the initiative programs, which aim to manage the lake's resources.

The largest of the Laurentian Great Lakes in North America, Lake Superior is bordered to the north by Ontario, Canada, and to the south by Minnesota, Wisconsin, and the upper peninsula of Michigan. With a freshwater volume of 2.9 million cubic miles and Lake Tanganyika at 4.5 million cubic miles, Lake Superior is only one-third the volume of Lake Baikal and Lake Tanganyika combined(Nelson, 2021). However, due to its vast surface area, Lake Superior is considered the largest lake in the world. Around 10,000 years ago, at the end of the Pleistocene period, when the last ice age ended, the land surface was damaged, resulting in the geologic development of Lake Superior. As a result, compared to Lake Baikal, which is among the oldest large lakes in the world, Lake Superior is among the youngest. At the end of the Pleistocene period, a similar process produced other Laurentian Great Lakes and the majority of North American lakes from temperate latitudes northward. About 200 rivers feed Lake Superior, which empties into Lake Huron via the St. Mary's River(Kauss, 1991). The number of fish species in Lake Superior is estimated to be 80. Because of the poor primary productivity and low nutrient content in the water, the lake is classified as oligotrophic. The total phosphorus level in mesotrophic lakes ranges from 12 to 24 mg/liter. These freshwater lake systems are thought to be comparatively healthy and somewhat productive. The Laurentian Great Lakes of North America include Lake Ontario, which is regarded as mesotrophic. The total phosphorus level of eutrophic lakes ranges from 24 to 96 mg/liter. Even though eutrophication can happen naturally, human activity such as farming or industry often has an impact on these freshwater lakes. Lake Ladoga has become eutrophic since the 1960s, most likely as a result of heavy manufacturing and farming activities in the watershed(Drabkova et al., 1996). The total phosphorus level in hypereutrophic lakes is higher than 96 mg/liter. These lakes are primarily oxygen-free toward the bottom, exhibit strong phytoplankton blooms, and have murky water. In hypereutrophic lakes, cyanobacteria like the Aphanizomenon flos-aquae seen in Upper Klamath Lake usually cause phytoplankton blooms. Plankton dynamics in limnetic zones of lakes, including phytoplankton and zooplankton, their dynamics and interactions, and their significance in the freshwater lake food web, are studied in limnology(Lemly & Dimmick, 1982). Higher aquatic plants play a crucial role in the littoral zone of freshwater lakes by providing food to a variety of aquatic organisms, including fish, ducks, and aquatic animals, as well as aquatic insects. Benthic organisms inhabit the lower parts of lakes, ranging from the coastal area to the deep oceans. Mollusks, annelids, crustaceans, and benthic fishes live in these settings.

1.12 WETLANDS

There are wetlands everywhere in the world. Wetlands make up 5–8% of the land area, with higher concentrations found in tropical (31%), subtropical (25%) and boreal (30%) regions and lower concentrations found in polar (2%) and subboreal (11) regions(Junk et al., 2013). Only 5% of wetlands are made up of saltwater, with freshwater making up the majority of them. The Pantanal, which spans 88,803 square miles across Brazil, Paraguay, and Bolivia, is sometimes cited as the largest wetland. Wetlands are frequently seen negatively as impediments to growth and progress or as disease reservoirs for mosquitoes that transmit diseases like malaria, and other diseases. Due to road construction, mining extraction, agriculture, the construction of residential and industrial complexes, and other human infrastructure, wetlands have been flooded and drained. Human activity has destroyed almost 50% of wetlands worldwide, and groundwater depletion and water pollution are affecting many more(Hu et al., 2017). Ninety percent of the wetlands have been lost in New Zealand, sixty percent in Europe and China, and fifty percent in Australia. The loss of wetlands occurs most quickly when a country is growing and developing. Wetland loss in developed countries has decreased as a result of new legislation protecting wetlands and the disappearance of wetlands that were formerly easy to drain and fill. Still, between 2004 and 2009, the United States lost 5,590 hectares per year. Because of national measures, the United States gained 89,140 hectares of wetlands between 1998 and 2004(Hansson et al., 2019). While more biologically significant vegetated wetland types continued to disappear, open-water wetlands frequently increased. Wetland loss rates in developing nations will probably be high as long as human population expansion and technology advancements persist.

The benefits that wetlands yield are disproportionately important when considering their relative size. People benefit from wetland ecological services, which include generating food and fiber, lowering flood damage, improving biodiversity, and offering a clean water source. Wetlands are very good at storing and cleaning water. They are frequently described as the kidneys of the landscape because of their capacity to eliminate toxins from the surrounding area. Because of their location, vegetation, and the many chemical reactions that take place throughout the alternating wetting and drying cycles, wetlands can reduce water velocity, which makes them efficient at eliminating pollutants and debris. In general, a system will work better the slower the water moves through it and the more intricate the flora and ground are. Wetlands absorb and store storm runoff, significantly lessening the effects of flooding. Their capacity to hold flood water lessens the force of water flows, which are most destructive to structures and other infrastructure. Instead, the water can be released gradually over a longer period of time. As a result, there is less damage from floods and less severe flooding. The greatest benefits for preventing flooding are often seen in wetlands near rivers. With net primary productivity of up to 1,200 grams per square meter, wetlands are the most productive ecosystem on Earth. Rich plant and animal life as well as intricate food webs are the outcomes of this high production. The most common residents of wetlands are waterfowl, which includes ducks, geese, swans, and other water birds including herons, egrets, and shorebirds. One of the most fertile regions in the world for producing waterfowl is the Prairie Pothole Region in North America. Because of the vast quantities of migrating ducks that the prairie potholes generate, it is referred to as the North American Duck Factory. For breeding birds, wetlands with an optimal proportion of open water to developing vegetation yield the highest productivity. Numerous species of animals, fish, amphibians, and invertebrates all find shelter in wetlands.

Many commercially useful goods are produced by wetlands. Peat is mostly harvested for the horticulture industry and, to a lesser extent, as fuel. The following fur-bearing mammals are mostly dependent on wetlands: muskrats, otters, mink, beavers, and nutria. The American alligator (*Alligator mississippiensis*), saltwater crocodile (*Crocodylus porosus*), and spectacled caiman (*Caiman crocodilus*) are harvested and used to produce belts and footwear. Wetland-dependent species comprise an estimated 95% of commercially caught fish and shellfish, such as pink shrimp (*Farfantepenaeus duorarum*) and various crabs, oysters, crayfish, salmon, catfish, and menhaden(Brotherton et al., 2020). In certain areas, frogs are also farmed for commercial purposes and their legs are taken. Bald cypress trees are producers of mulch and commercially valuable timber used in construction. Many other trees, including swamp white oak (*Quercus bicolor*), and pecans (*Carya illinoinensis*), are also important for their timber, while pecan trees yield valuable nut crops. Numerous herbaceous marsh plants, including paper reed (*Cyperus papyrus*), common reed, and cattails, are productive enough to yield materials that can be used to make mats, baskets, and even roofing. In addition to being collected for human food, wild rice is also frequently packaged and sold for high rates, especially by indigenous communities. Northern wetland systems and upland areas that have been turned into wetlands, especially for cranberry cultivation are the usual places to cultivate cranberries(Hoekstra et al., 2020). Political bodies, including the US government, have enacted laws and regulations to conserve wetlands because of the important ecosystem services and functions that many wetlands offer.

Functional assessment mitigation is becoming more and more common, while size-based mitigation has historically accounted for the majority of mitigation efforts. Larger wetlands must be constructed in order to compensate for the loss of more complex vegetated wetland types caused by a size-based mitigation strategy.

Therefore, compared to developing wetlands, forested wetlands require a higher ratio of created to destroyed land, and both require more than open-water wetlands. The recommended method of compensating mitigation for inevitable wetland losses is through mitigation banks. Companies that are prepared to commit to long-term investments to create wetlands before it is destroyed are typically those who construct wetland banks. For their own use, large public entities may also create wetland banks. Larger wetland sites offer more ecological value; mitigation happens before impacts so functional success is already known; and project review time is shortened, which shortens the time it takes to obtain permits and improves the efficiency of regulatory agencies. All of these benefits make mitigation banking more cost-effective than traditional compensatory mitigation.

Apart from specific state regulations linked to wetlands, these isolated wetlands are now unprotected. To preserve the ecological integrity of wetlands of global significance, member nations of the Ramsar Convention were signed. Although the Ramsar definition is more expansive than other wetland definitions, it is nonetheless a useful tool for protecting wetlands across international borders and gives wetlands in partner countries more prominence. With 1,960 sites covering 190 million hectares of internationally important wetlands, 160 countries are parties to the treaty(Smardon, 2009).

1.13 DESERTS

Deserts are typically arid regions. There are some exceptions: For example, the majority of the Kalahari Desert is semiarid, whereas coastal deserts can be semiarid or dry semi-humid during specific seasons or in specific regions. Coastal deserts are one of the three main types of deserts based on climate. Coastal deserts are defined as those that frequently experience fog due to their proximity to or location on a coastline. Cold deserts, such as those found in central Asia, and hot deserts, such as those found in Australia, are classified according to their average temperatures. As distinctive to the coastal desert ecology as the summer highs and winter lows of hot and cold deserts are these fogs, which bring moisture to an otherwise dry region. Many coastal deserts have more cloudy days than wet ones; this fog could be the primary supply of moisture in some areas. Due to the ocean's moderating influence, coastal deserts also have fewer fluctuations in temperature. Coastal deserts typically have a diurnal range of only 50 degrees, comparable to less-arid regions. This is in contrast to hot and cold deserts, which frequently experience a diurnal range of 95 degrees F or the difference between cold night-time and hot day-time temperatures. Even if it is simpler to characterize deserts in terms of their vegetation and soils, climate has an impact on, if not directly determines, soil and vegetation. This is because there is a lack of permanent human settlements in deserts, which has resulted in a scarcity of climatic data. Mineral-dominated desert soils have minimal organic matter. They often have good drainage, a fine texture, and are porous. To benefit from storms, coastal desert plants often have large root systems, leaves and stems that store water, and occasionally ridges and grooves that enable stems to swell when they are filled with water. Common coastal desert flora includes rice grass, black sage, saltbush, and buckwheat bush.

In the desert, moisture has a complicated role since it greatly influences the ecosystem and the landscape. Plant life depends on the constant availability of water. In coastal desert fog zones, even plant life has evolved specifically to survive in the fog, unlike other plant life that is dependent on intermittent rainfall. There are fewer grasses and root systems to hold the soil in place, as well as less microbial activity and organic matter breakdown influencing the soil's composition. Because the landscape has fewer barriers against erosion, rain has a greater erosive effect when it does occur. For example, an annual rainfall of 12 inches (30 centimeters) may have a greater effect than four or five times that amount in another environment. Furthermore, a single rainstorm has the power to completely alter the topography of the land and destroy delicate ecosystems. The efficiency of temperature extremes is enhanced by the presence of even trace moisture, which aids in the weathering and fragmentation of rocks. Additionally, in coastal deserts where these extremes are less frequent, the precipitation leaves crystal deposits in the soil, which can fracture rocks as they grow. Because there is less rainfall to wash it away, salt is more prevalent in the soil even in non-coastal deserts than in non-desert areas. Sodium chloride and gypsum, which are more likely to occur in fog zones or semi-arid environments; and calcrete, a duricrust rich in calcium carbonate, are examples of coastal desert soil salts. The sand rose is a collection of flat gypsum crystals that resemble rose petals. In addition to ordinary salt and gypsum, the coastal Atacama Desert contains rich quantities of perchlorates, iodates, and nitrates that have been mined for fertilizer since the 1830s(Soledad Ureta-Zañartu, 2022). The biofilm, or thin coating of organisms that develop on the surfaces of rocks, in the cracks, and in their pore spaces, is an essential component of the coastal desert environment. These species include fungi, bacteria, and cryptogams, which are organisms that reproduce by spores and are most typically seen in deserts as lichen, algae, and mosses. Through the acids it excretes, the biofilm not only creates a living environment inside the pores and crevices in rocks, but it also plays a vital role in arid areas and the disintegration of rock into distinctive desert soils.

In many situations, but particularly in deserts, where a thin layer of clay, manganese, and iron paints the surface of exposed rocks, turning them brown or black, microbes are also involved in the development of desert varnish. In the fog zones of coastal deserts, where cryptogams predominate and can better utilize moisture than vascular plants, microbes also flourish. In order to understand more about the kinds of life that desert fog water can support, researchers have examined the chemical composition and ion concentrations of the water. They have discovered that while the water salinity, acidity, and levels of certain chemicals, like calcium, nitrous oxide, and manganese, can occasionally exceed the recommended levels, these levels are still within WHO guidelines for collecting water for human consumption. To put it another way, they are safe for the majority of animal life; the main differences arise from the way moisture is collected, and the way trace elements gradually accumulate over time, like salt building up in the soil. A large portion of the fauna in coastal desert fog zones gets its water from the food they eat. Additionally, some species have evolved to get their water from the fog. There are thirteen distinct *Onymacris* species in the Namib Desert(Mitchell et al., 2020). They are able to escape the heated layer of the desert sand because they have longer legs than other beetles, which resemble spiders. Their method of getting moisture is called fog basking. Two species in particular, *Onymacris unguicularis* and *O. bicolor*, have been seen flocking to the tops of dunes when fogs roll in over the coastal desert of Namibia (Huntley, 2023). This is so that the fog can condense on their abdomens, which are colder than the sand due to their long legs. It has been demonstrated that fog basking is an adaptation to desert circumstances.

The presence of the ocean nearby may have an effect on the wildlife in a coastal desert in addition to fog. The black widow spider, *Latrodectus indistinctus*, is a thriving species in the Namib Desert. While its main effect is to decrease the number of herbivores on the dunes, protecting the dune vegetation and the surrounding microbial ecosystem, its high population is not sustained by prey found in the desert, but rather by flies that have migrated into the area from the nearby marine biome and feed on detrital algae. Coastal spiders are six times more common in Baja, California than they are inland, and their primary food source is the sea. Studies of coastal desert mammal populations indicate a strong reliance on marine species, which make up more than half of the diet of the coastal coyote. The Baja desert system is also home to 3–24 times more insects, scorpions, lizards, rats, and coyotes on the coast than inland. Coastal rodent populations are also negatively impacted by coyotes; on desert islands off the coast of Baja, coyote-free islands have far fewer rodents than coyote-rich islands. The kinds of algae that are present in the ecosystem vary as a result of the ocean's proximity. Many invertebrates in the Atacama Desert feed on the marine green alga Ulva, which is then eaten by geckos (*Phyllodactus angustidigitus*), solifuges (*Chinchippus peruvianus*), and scorpions (*Brachistosternus ehrenbergii*)(Catenazzi & A. Donnelly, 2007). Because part of the desert's fauna will eventually be preyed upon by predators, and so on as those nutrients migrate further inland as they pass down the food chain, coastal deserts are essential for the transfer of nutrients from marine biomes to terrestrial biomes. It has also been shown that fertilization from dust high in phosphorus that is blown in from the coastal western Sahara Desert, 3,107 miles away, is necessary to maintain the productivity of the Amazon rainforests. Because there was no rain in the Atacama Desert for more than 400 years, from 1570 to 1971, it is said to be the driest desert in the world. Even if floods occur frequently in other coastal deserts, they are not always able to take advantage of the moisture that becomes available quickly. When flooding occurs in certain deserts, temporary lakes may form; if this occurs frequently enough, an ecosystem may grow around the lakes. On the other hand, marine life will accompany floodwaters that originate from overflowing rivers or other sources of water outside of the desert environment. In certain instances, they may also introduce new microbial life or other fauna, or they may draw migratory birds or other fauna. The introduction of a new species can have disastrous effects in the desert due to its limited life. In Australia, one such issue is the overgrazing of newly imported rabbits, which threatens the grasses that stabilize the desert dune system.

Because these species depend heavily on the balance of their particular ecosystem, endemic species are frequently well-adapted to it; however, unlike species that can thrive in a wide range of environments, they may not be as resilient to new threats like disease, predators, or resource competitors as more common species are, having successfully navigated these obstacles many times during their evolutionary history. When it rains, there is a surge in biological activity; some of this energy is consumed, while the rest is stored as seeds or as water in the form of succulents or other vegetative life. In the Arctic, for instance, different microflora and microfauna may almost stop throughout the winter when the sea is covered in ice and there is little access to sunshine or fresh nutrients, only to recommence after the summer thawing. Many animals in the desert store up their energy throughout the dry seasons so they may react quickly when rainwater suddenly becomes available. While smaller rodents are adapted to move very slowly in order to use less energy, kangaroos are adapted to travel large distances from one food source to another in the most energy-efficient manner possible, even within the same desert. In coastal deserts, certain amphibians have accelerated larval stages that enable them to reach maturity sooner and increase their chances of survival; on the other hand, some toads that live in burrows effectively hibernate during the dry season, blocking their burrows with gelatinous slime that is washed away by the subsequent wet season.

1.14 MORE ON GLOBAL DESERTS

Although fog can provide some moisture to coastal deserts, this effect is limited to a small area of the desert, and the pulse-reserve paradigm still governs how the desert reacts to precipitation. Pulses replenish soil water more effectively than fog, which promotes plant development and may even cause some animals to mate. Because the root systems of shrubs and grasses form macropores in the soil that allow more water to penetrate, these ecosystems make the best use of floodwaters. Eventually, the water will evaporate or flow into a different ecosystem at a lower elevation. Different C_4 photosynthesis types in plants in the Namib Desert react differently to rainfall(Cabido et al., 2008). These plants all fix carbon dioxide (CO_2) in the mesophyll cells to create oxaloacetate. With an inner bundle positioned between the Kranz sheath and the metaxylem components, some convert the CO_2 primarily into aspartate. Others use a single chlorenchymatous or Kranz sheath and centrifugal chloroplasts developed around vascular bundles to convert CO_2 predominantly into malate. As rainfall increases, so does the abundance of malate converters, whilst aspartate decreases. The malate converters have less developed grana and a higher mitochondrial frequency than the aspartate converters. Deserts at mid-latitudes are defined by arid conditions and unique physical and geographic characteristics. Many species have adapted to survive in this extreme environment. Some well-known examples include plants that have changed how they photosynthesize to conserve water during the hottest parts of the day and animals that have modified their behavior to regulate their body temperature. Some main risks to this ecosystem are overuse of resources, desertification, erosion, and urbanization. Midlatitude deserts can be initially misleading due to their appearance of dryness and lack of life. In terms of primary productivity, desert ecosystems are weak in comparison to temperate forest ecosystems throughout geologic timescales. Nonetheless, mid-latitude deserts are packed hotspots of life, frequently harboring organisms that have evolved to become adaptable to these hostile conditions. A wide variety of humans that have similarly adapted to the desert climate live in mid-latitude deserts. The diversity of desert inhabitants is best demonstrated by their adaptations to their environment and their protection systems; examples include the spines typically seen on cacti and the distinctive defense mechanisms used by lizards.

Mid-latitudes deserts are the most prevalent kind of deserts, such as the Sonoran Desert in North America, the Great Sandy Desert in Australia, and the Sahara in Africa(Hereford et al., 2006). After the dry air in the tropical latitudes has emptied moisture, it circulates down to the ground, creating mid-latitude deserts. The tropical latitudes receive the maximum solar radiation, which causes warm, humid air to rise to the atmosphere. Nearly all of the precipitation in the air rises to the wet tropics, causing dry, chilly air to circulate down from the atmosphere at a latitude of about 30 degrees north or south of the equator. The high levels of aridity and precipitation patterns found in the majority of mid-latitude deserts are explained by this dry air. The precipitation-to-evaporation ratio of a desert can be used to describe most of them. Evaporation usually exceeds the total precipitation that falls on the ground in most mid-latitude deserts. Given their global distribution and the broad definition of mid-latitude desert, the overall amount of precipitation received by these deserts varies considerably. With an average yearly precipitation of less than 9.84 inches (25 cm), the majority of these deserts are exceedingly dry. In most of these deserts, there is minimal cloud cover due to the less precipitation. Wide variations in daily temperature are caused by low cloud cover. In the worst circumstances, daytime highs of 104 degrees Fahrenheit (40 degrees Celsius) and nighttime lows of about 32 °F (0°C) are possible. The cool, dry air that frequently enters mid-latitude deserts can become extremely violent, creating large sandstorms with winds of up to 37–62 miles per hour (60–100 kilometers per hour) and dust that can reach as high as 1.5 kilometers. This kind of habitat is characterized by the unique geophysical characteristics of mid-latitude deserts. Large expanses of sand dunes are common in mid-latitude deserts; the western Sahara Desert is home to some of the world's largest and tallest dunes, which can reach heights of around 1,000 meters(Goudie et al., 2021; Hamdan et al., 2016). Large expanses of salt flats or dry, rocky outcrops cover most of the mid-latitude deserts. Salar de Uyuni, located in the northernmost part of the Atacama Desert, is the largest of these salt flats, covering an area of around 4,054 square miles. While they are uncommon in mid-latitude deserts, some rivers like the Nile cross a significant amount of the Sahara. Rivers that flow across mid-latitude deserts typically have their source in higher-lying areas with higher precipitation levels.

Oasis, which are normally created by subterranean river systems or aquifers, are another common source of freshwater in mid-latitude deserts. The majority of the diversity of species in mid-latitude deserts can be found in both rivers and oases. The majority of deserts in mid-latitudes are very young ecosystems, having formed fewer than 20,000 years ago. The glaciers that had covered the northern and southern latitudes started to recede, and historically, temperate regions experienced desertification. As a result, mid-latitude deserts expanded in size. Nevertheless, despite being relatively young in geologic and evolutionary timescales, numerous unique organisms with attractive evolutionary histories can be found in mid-latitude deserts. The difficult environmental circumstances of restricted water availability, high temperatures, scarce food supplies, intense competition, and high predation have led to the evolution of features that enable mid-latitude desert species to thrive. Mid-latitude deserts are intriguing locations for evolutionary biologists to study evolutionary phenomena like convergent evolution because of the evolution of life in the desert. When comparable traits arise in species without a common ancestor, it is known as convergence. This is primarily due to shared selection pressures across many locations. The Australian mid-latitude desert Hopping mice and the North American mid-latitude desert kangaroo rats are two examples that have acquired convergent features in these environments. Succulent mid-latitude desert plants have also evolved convergent features, such as the spines found in African Euphorbia species and North American cactus, to ward off herbivory. Water availability and the physiological and behavioral adaptations of species to extreme temperature changes are critical for life in mid-latitude deserts. There are just a few primary water sources, such as oases, rivers, and aquifers, in mid-latitude deserts. The majority of the biodiversity in mid-latitude deserts is located along these riparian zones. Nevertheless, several organisms have evolved to extreme aridity by obtaining water from fog as exemplified by Euphorbia found in Atacama Desert, the world's driest mid-latitude desert. Additionally, species especially vertebrates have evolved their behavior to adapt to the extreme temperature swings found in mid-latitude deserts. Desert species are nocturnal in nature. Deserts with abundant endemic biodiversity are the product of evolution and adaptation. The most biodiverse mid-latitude desert in North America is the Sonoran Desert(Ezcurra et al., 2020). Urbanization, water resource depletion, and habitat disturbance pose threats to biodiversity in mid-latitude deserts. As people continue to occupy the mid-latitude deserts, water supplies are being exhausted, overgrazing is altering the composition of the desert plant and animal species, and towns are growing in unsustainable places for human habitation. To properly comprehend how these disturbances affect the native species of the mid-latitude deserts, more research is necessary. The largest terrestrial oil resources are found in mid-latitude deserts. Among the world's largest oil deposits are found in the mid-latitude deserts of northern Africa and the Middle East. Six of the ten biggest oil-producing countries, Iraq, Kuwait, the United Arab Emirates, Iran, Saudi Arabia, and Libya are located in these arid regions. Mid-latitude deserts are crucial locations for the extraction of these fossil fuels for transportation and energy production. The deserts of Iraq and Kuwait have been affected by oil drilling and poor oil production management, particularly as a result of recent wars and political unrest. Several millions of barrels of oil spilled into the desert sands during the Gulf War in the early 1990s destroyed thousands of acres of vegetation and fauna in Kuwait. There is still work being done on restoration. To save the endemic and endangered species that live in certain regions of the planet, oil production and regulation in mid-latitude deserts should be closely watched over and managed. Mid-latitude deserts are rich in mineral resources. In the driest parts of mid-latitude deserts, minerals leached from groundwater and rainfall often accumulate due to high rates of evaporation. In many mid-latitude deserts, salts commonly accumulate close to the surface. The Atacama Desert in South America is one of the mid-latitude deserts that produces the salt. Gypsum and boron are two more evaporites, or mineral deposits that collect in mid-latitude deserts after evaporation removes the elements from solution. Additionally, gypsum accumulates underground, which accounts for the enormous crystals found in the extensive cave systems found in the deserts of southwestern United States (Lechuguilla Cave). The soils of mid-latitude deserts frequently accumulate significant amounts of important metals. Copper mining in the Sonoran and Chihuahuan deserts is an important industry in North America. Lead, zinc, and iron are additional metals that are present in midlatitude deserts. Many of the midlatitude desert regions are also regularly mined for economically valuable metals, such as uranium, silver, and gold(L. Tang & Werner, 2023).

Mineral extraction, processing, and management usually have a negative impact on the fragile desert ecosystems. Large swaths of the American environment are altered by open-pit copper mining, particularly in the desert states of Arizona and New Mexico. These places are unlikely to recover from such a substantial disturbance. Refineries' exploitation of these resources damages desert ecosystems by contaminating soils with heavy metals and causing metals like lead, copper, and zinc to bioaccumulate in the plants and animals of the desert. It is common to find heavy metal pollution and bioaccumulation up to 40 miles away from a refinery. Mid-latitude deserts offer numerous advantages and natural resources that can be extracted with little negative ecological impact. Hydroelectric power has been produced by regulating the large rivers that cross several mid-latitude deserts. Although the desert ecosystems are damaged by the damming of rivers to generate hydroelectric power, the sustainability and overall ecological effects of energy generation from these sources may be superior to that of burning fossil fuels. Only approximately 0.1 percent of energy is produced globally by dams in mid-latitude deserts, compared to about 2 percent by hydroelectricity(Capellán-Pérez et al., 2017). On the other hand, mid-latitude deserts are perfect for the advancement of wind and solar energy generation. Mid-latitude deserts have the potential to develop into important global providers of sustainable energy production, and these energy sources have the capacity to supply about 33% of the world's energy needs. The Mojave Desert in the United States is already home to several large-scale solar energy production projects. The greatest potential for producing solar energy is found in the deserts of the mid-latitudes. The Sahara, Arabian Peninsula and the deserts of Australia are excellent locations for solar energy collection. The mid-latitude desert geomorphology makes them ideal for the growth of wind energy production. In the US, desert regions in Texas, California, and New Mexico are already being developed for the production of substantial amounts of wind energy. As of now, 2.5 percent of the energy produced in the U.S. comes from wind energy; but, with further development, by 2030, that percentage might reach up to 20 percent. The ancient Mesopotamian civilizations that grew along the productive river basins in the mid-latitude arid regions in the Middle East are among the most famous. The Tigris, Euphrates, and Nile provided the riparian habitat that was the center of ancient Mesopotamia, but the harsh mid-latitude desert environment forced the people who formed these civilizations to adapt.

Chapter 2

IMPORTANCE OF WETLANDS

Contents

2.1 INTRODUCTION

Water is one of the few substances that can exist in the range of temperatures. Water is frequently observed in its liquid condition, which is found between 32°F and 212°F (0°C and 100°C), with the exception of high mountains and the polar regions. While liquid water completely turns into vapor above its boiling point, some water still exists in this state at lower temperatures. Water enters the atmosphere through the surfaces of all bodies of water, such as lakes, rivers, streams, vegetation, and soils. When compared to bare soil, vegetation produces comparatively high amounts of water vapor. This is due to the fact that as leaves photosynthesize, they inhale carbon dioxide from the atmosphere through microscopic pores on their surface known as stomata. Transpiration is the mechanism by which the leaf pores release water vapor(Rodov, 2002). Hence, by transpiration or evaporation, water vapor is released into the atmosphere from all land and aquatic surfaces.

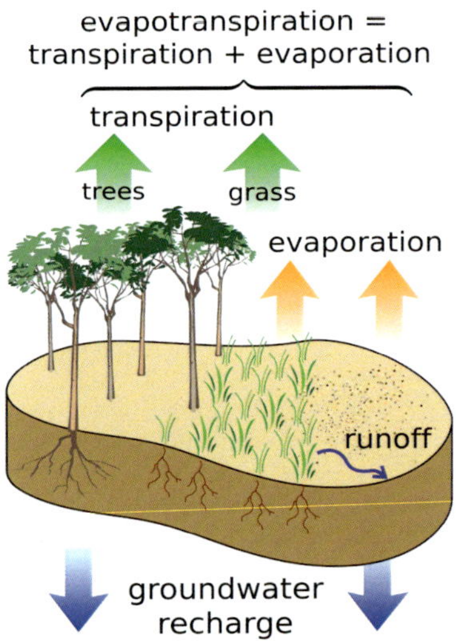

Figure 2.1 Evapotranspiration

Source: https://commons.wikimedia.org/wiki/File:Surface_water_cycle.svg,*Creative Commons License.*

Evapotranspiration (Figure 2.1) is the term used to describe the combination of these two sources of water vapor. When air cools, it may store less water vapor than hot air. Hot air can hold more vapor than cool air. As a result, droplets of water condense and create clouds. These droplets turn into rain when they get larger. Snowflakes are formed when water droplets turn into ice crystals when the air temperature falls below that of water(Geresdi et al., 2014). Precipitation is the term used to describe water falling down from the atmosphere as either raindrops or snow. Snow compacts into ice sustained for a very long time when it falls in areas that are cold, like polar ice caps or extremely high mountains. The earth is supplied with liquid water by rainfall and melting ice, which follows gravity pull and flows through wetlands, over rocks in mountain streams, soaks into the soil, and drains through porous rocks. The hydrological cycle is the name given to the practically constant movement of water.

2.2 OCEANS-THEIR CONNECT TO WETLANDS

Oceans comprise 97% of the surface water, which is saline. 2.25 percent of the remaining 3 percent, which is made up of freshwater, is trapped in the ice caps and glaciers. This may seem like a relatively small amount of the water resources in the globe, yet it is a crucial part of the water cycle. Each species that inhabits the earth depends on it for survival; they all require a daily quantity of this essential substance. The seas contribute around 84% of the total amount of water vapor that enters the atmosphere, with evapotranspiration from the land surface, including lakes, marshes, and vegetation supplying the remaining portion. The flow of water over the surface of the land as it makes its way back to the ocean shapes the landscape by eroding the materials in its route and forming canyons and river valleys. Water flows through rocks and soils, dissolving chemical components that are then transferred to the sea. However, when water evaporates again, these constituents are generally lost, making the seas more and more rich in chemicals and salts. With a few exceptions, wetlands found inland often have low quantities of various elements because water flows through them continuously, preventing pollutants from building up. However, lakes, wetlands, and peatlands slow down the flow of water, lessening the erosive effect of the flowing water and serving as temporary storage areas. Anywhere there is a consistent supply of precipitation or drainage water, wetlands can develop. Wetlands are not constrained to specific climatic zones of the land surface, in contrast to many other biomes like tropical rain forests, savannas, deserts, temperate forests, or tundra. Wetlands are common in areas with high precipitation rates, away from the coast, where there is a steady supply of water. Tropics have a lot of rainfall, and because the northern regions are often cooler than the southern ones, there is less transpiration and evaporation, which means that more water stays in the soils.

The Northern Hemisphere has more wetlands than the Southern Hemisphere. The majority of the wetlands in the Northern Hemisphere are found between latitudes 45°N and 75°N; in contrast, the Southern Hemisphere has very less land(Aselmann & Crutzen, 1989). Most of the year, the sun shines nearly directly overhead at noon, delivering solar radiation to the equatorial areas. More energy reaches the ground when light falls vertically because it travels through a thinner atmosphere than it does in temperate zones. Higher latitudes, or those nearer the poles, receive energy from the sun at a lower angle than those at the equator, allowing energy to be dispersed across a larger area. Because light penetrates deeper into the atmosphere, more energy is absorbed. The equatorial regions get warmer than the temperate zone as a result of these variations in radiation. Denser, colder air traveling toward the equator pushes the warmer, moister air of the tropics higher. Air cools as it rises and air temperature decreases by around 3.6°F, or 6.5°C, for every 1,000 feet we ascend above the surface. Water droplets condense in the rising air and fall as heavy rain in the equatorial zone because cooler air can carry less water vapor than warm air.

High altitude produces cool air, which is forced away from the equator and begins to sink again because it is denser than warm air. This cool, dry air lowers and creates an area with an arid environment between latitudes 20° and 30° N and S of the equator(Dirmeyer & Shukla, 1996). Wetlands are rare and deserts predominate in the Northern and Southern Hemispheres. The wetlands that are there are either found along the coast or, as is the case in central Australia, may emerge due to heat and drought. In these latitudes, the air is dense, spreading outward toward the poles and the equator. Eventually, masses of dense, cold air from the poles traveling toward the equator clash with the poleward-moving mass of warmed air. As the denser polar air mass presses beneath it, the warmer air that has traveled poleward and gathered up moisture is forced upward by this collision, greatly increasing atmospheric instability. In this unstable area i.e., between latitudes 50° and 70°, there is a lot of precipitation, which precipitates as snow in the winter and rain in the summer. Therefore, the global pattern of precipitation coming from the unequal heating of the surface and the ensuing air movements can explain the overall pattern of wetland zones. Summer brings the hot air closer to India, which is located within the tropics but north of the equator. As a result, rainfall increases in the form of the monsoon, which is known as heavy rain. The high Himalayan mountains, which form a chain along the northern edge of India, block the rains from moving farther north into Asia. As a result, the precipitation that is trapped by the mountains comes back into India, Pakistan, and Bangladesh, creating wetlands in these low-lying areas. Rain-bearing air masses are unable to penetrate to the north, beyond the Himalayas, where desert conditions are the norm.

2.3 OCEAN CURRENTS AND WETLANDS

Ocean currents have an impact on precipitation patterns and, consequently, the distribution of wetlands. The North Pacific drift is a warm oceanic current that travels east across the Pacific Ocean, passing north along the coast of British Columbia in Canada before reaching the west coast of Oregon and Washington in the United States(Tabata, 1975). This current creates warm, humid air that brings a lot of rain to the shoreline, which leads to the growth of wetlands and temperate rainforests. Similar warm currents, known as the Gulf Stream, flow eastward across the North Atlantic Ocean, carrying water from the Caribbean Sea to the western portions of northern Europe. This phenomenon results in heavy precipitation and the extensive growth of wetlands in these areas. Therefore, the overall pattern of wetland distribution across the surface can be explained by the combined effects of solar energy intake, air mass movements, the influence of high mountains, and ocean currents. We need to look at the patterns of wetlands and the forces that influence them at the landscape scale.

Figure 2.2 Wetlands

Source: Vymazal, J. Constructed Wetlands for Wastewater Treatment. Water 2010, 2, 530-549. https://doi.org/10.3390/w2030530,*Creative Commons License.*

Wetlands (Figure 2.2) are more likely to be found in the lower areas of valleys and floodplains because water flows downward under the pull of gravity and moves fast where there are steep slopes. As the slope grows less steep, water not only moves more slowly but also accumulates in the valleys from the surrounding areas. The valley floors can be thought of as the receiving areas, and the hills as the land from which water is collected. The catchment or watershed of a valley is the area where water collects. The amount of water that enters the valley is significantly influenced by the size of the watershed in comparison to the receiving area of the valley floor. Depending on whether part or all of this water is held there or instead flows quickly through the valley and toward the sea, a wetland may or may not occur in such a valley. Wetland development will be influenced by the bedrock type, the slope of the valley floor, and the existence of any obstructions preventing water from flowing out of the valley.

It is believed that all of the water entering a catchment will flow into the valley. Not all of the precipitation, particularly if the watershed's slopes are covered in forest, reaches the ground when it arrives as rain or snow. A portion of the rain or snow that is intercepted by the vegetation canopy re-evaporates when the precipitation stops, never into the soil. Thus, vegetation functions somewhat like an umbrella, keeping part of the rainfall from getting to the areas underneath it. However, there are instances in which vegetation might behave differently and bring more water into a catchment. Water droplets condense on the leaves and twigs of the trees and drip to the ground when mist and fog cover a forested mountain slope, increasing the water supply. This is known to meteorologists as occult precipitation. It may return to the atmosphere as evaporated, or it could be absorbed by the soil and stay there due to the soil organic debris, which includes dead leaves and wood, acting as a sponge. The roots of the developing plants absorb some of the water from the soil. Whatever is left could eventually flow into the streams or wetlands as a result of gravity. The type of vegetation in a watershed has a significant impact on how much water flows into the valley. Many studies have been carried out by ecologists in the Appalachian Mountains of the eastern United States, changing the flora in a catchment to see how it affects water movement. The amount of water that flows into a valley can rise by 10 to 40 percent when the temperate deciduous forest is cleared from a watershed. This is a result of the removal of the canopy interception and plant transpiration requirements. However, there is a 10–20% decrease in the amount of water moving to the valley if oak forest is substituted with pine forest(Lovett & Goodale, 2011; Aguadé et al., 2015). This is because pine requires more water than oak does, in part due to its high transpiration rate and also because pine is evergreen, meaning that it can absorb and evaporate water even in the winter.

2.4 VEGETATION

The vegetation in a watershed greatly influences water movements, the surrounding vegetation and land use also influence whether or not a wetland develops. Even though they are prevalent in low-lying areas that receive water from large watersheds, wetlands can grow on sloping terrain if there is enough precipitation. Located in the northeastern region of North America, Newfoundland experiences year-round cold temperatures and a lot of rainfall. This indicates that water soaks into the surface continuously rather than evaporating fast. Because of these circumstances, wetlands can develop even in improbable places, including slightly sloping hills where a steady supply of water causes saturation even when gravity causes the surface water to drain. Blanket bogs are wetlands that produce peat and can cover hills and valleys, creating a blanket-like effect on the surrounding area. They can be found in northwest Europe, on a few islands in the Southern Hemisphere, and even on high mountains near the equator, where a combination of cold temperatures and heavy precipitation prevails. These unique wetlands serve as an example of how the distribution pattern of wetlands in the landscape can occasionally surprise us.

Figure 2.3 Freshwater lakes

Source: Yücel, A.; Markovic, M.; Atilgan, A.; Rolbiecki, R.; Ertop, H.; Jagosz, B.; Ptach, W.; Łangowski, A.; Jakubowski, T. Investigation of Annual Lake Water Levels and Water Volumes with Şen Innovation and Mann-Kendall Rank Correlation Trend Tests: Example of Lake Eğirdir, Turkey. Water 2022, 14, 2374. https://doi.org/10.3390/w14152374, *Creative Commons License.*

Freshwater lakes (Figure 2.3) and ponds are found in all segments of the earth, from the Arctic tundra to the equatorial regions. They can even occur in desert regions of the planet, such as Lake Chad in the southern Sahara, however, during dry spells, water levels in these areas frequently fluctuate. Although freshwater lakes in such diverse temperatures have distinct kinds of vegetation and animals living in them, they nonetheless share several essential characteristics. The only types of plant life found in water deeper than 13 feet are usually microscopic plankton, or free-floating aquatic plants that don't require roots. Only shallower water is suitable for the growth of plants with floating leaves that take root at the bottom, like water lilies. Animals can live in both shallow and deeper waters, ranging in size from microscopic zooplankton to much larger fish, mammals, or birds. More plant species can develop on the buried portion of soil as lake borders get shallower due to sunlight penetrating the bottom; however, this is partially dependent on how turbid, or cloudy, the waters are. Silts, clays, and organic matter are examples of suspended materials that can heavily clog waterways, making it difficult for light to penetrate and restricting the growth of bottom-dwelling plants to relatively shallow depths. The line separating the lake muck from the water above may be covered in thin filaments of green algae. There are tall plants that may grow stems that reach above the water surface by rooting in the sewage. Above the water surface, there is a mass of flora, mostly in the shape of vertical stems, in addition to the complex pattern of entwined roots beneath the surface. A habitat's biodiversity increases with its structural complexity. Hidden from their rivals among the tall reeds, birds like bitterns and herons can hunt fish and amphibians. Certain birds, like the European reed warbler (*Acrocephalus scirpaceus*), build nests that are suspended above the water surface and swing amid the vertical reed stalks. Marshes can form when floating mats of stems and roots penetrate shallow waterways, although they are typically formed by developing plants rooted in the base(Sasser et al., 1995).

2.5 THE INTEGRAL ROLE OF FENS

Wave activity from storms frequently affects the marshes of Lake Victoria and other lakes in East Africa. Although agricultural drainage and urbanization have caused the marshlands in the temperate regions of North America and Europe to become smaller and more dispersed, they nevertheless serve as a home for animals. Fens have a water table below the surface of their soils. The soils are always moist to the touch and they are still wetland ecosystems, but in the summer, one needs to dig a hole to determine the water level, which can be up to 30 cm (or a foot) below the surface. Since they are rheotrophic ecosystems, water enters them from catchments and passes through them before entering streams. They are submerged in water throughout the winter and during floods, but as the season progresses, they progressively dry up. A fen often has a more varied vegetation than a marsh, which is mostly made up of one or two cattails or reeds. Instead, the vegetation of fens is shorter and typically has a range of flowering plants than the homogeneous marsh with its tall and prominent reeds. Air seeps into the soil during the dry season, providing a growing medium for plants unable to withstand the marsh flooding. But unlike marshes, fens have a very limited bird population because the flora does not offer bitterns enough cover or suitable nesting sites for birds that depend on stalks to support their nests. However, there are still plenty of amphibians in the fens, where egrets and herons eat them.

The fen serves as a food supply for several grazing animals, including deer, due to its shallow water in the winter and its drier surface in the summer(Vernon et al., 2022). Fens support diverse forms because they rely on their watersheds for both chemical and water components; the vegetation's availability of chemical nutrients is influenced by the geology of the catchment. Rich fens are those that are overflowing with the nutritional elements required for plant growth. The intensity of the mineral supply has an impact on the diversity of life. Fens can be found globally, wherever a catchment produces water that collects and flows. Damp hollows can create a fen even in the dry conditions of sand dunes, both coastal and inland. Known as dune slacks, these unique types of fens are diverse in terms of both plant and animal life. Seepage of seawater into the coastal dune slacks can create a slightly saline, or brackish, habitat that is home to a specific spectrum of organisms. Fens can be very large and have different surface patterns in cool northern sites. In Scandinavia, Russia, or Canada, we can see valley fens scattered over their surfaces. These are known as string bogs in North America, although they are known as aapa fens in Europe. The string bogs are actually fens. Raised ridges that are three feet above the water table and follow the contours of the fen's slope are what give rise to the stripes. There's a chance that these ridges are dry enough for trees like black spruce to grow on them. The low-lying lakes that divide the ridges are frequently covered in bog mosses (Sphagnum). Because of the variations in flora and the open water, the color contrast between the ridges and pools is quite noticeable from an aerial perspective. Water meanders around the pools as it descends the wetland slope, occasionally slicing through ridge gaps to reach the low-lying areas.

2.6 WETLAND ECOSYSTEMS

Rheotrophic fen is a different kind of fen ecosystem that occurs in mountainous areas with steep slopes and water seeping from the ground due to outcrops of water-resistant, impermeable rocks(Chakraborty et al., 2023). Springs are frequently the source of rivers and streams. However, the water may flow over the surface and form a small wetland that is abundant in both plant and animal life before it erodes into the rock and soil. High calcium carbonate (lime) concentrations in the water can cause this compound to create vast lime deposits. Among the tufa deposits, plants develop and produce a complex mixture of lime and organic peat that can expand. Spring mires can be found in belts of coniferous or deciduous woodland, or among alpine plants on high mountains. Their ability to periodically erode can result in unstable soils, which is one of the characteristics that attract environmentalists. After the glaciers receded around 20,000 years ago in regions that included Canada, some areas of the northern United States, northern Europe, and parts of northern Asia, the climate warmed up and the glaciers withdrew, causing the forest to grow. Because of this, a large number of alpine plants and invertebrate species have disappeared from the southern regions. These small areas of moist, eroding soils have produced the ideal environment for the local survival of the alpines. In addition to being rich in species, spring mires frequently have uncommon mollusks, insects, and plants that have dispersed.

The vegetation in fens and marshes is non woody and herbaceous. In contrast, trees predominate in marshlands. Very few trees can survive in highly moist soils, like the Florida bald cypress and the northern tamarack. Trees may eventually spread into drier areas such as fens and marshes. As wetlands grow, alders (Alnus) and willows (Salix) are especially effective at creating wet forests, an ecosystem known as carr in Europe. As is the situation in the southeast of North America, wetland forests can also be found in the low-lying floodplains of rivers. While the Big Cypress Swamp of southern Florida covers 1,500 square miles and is dominated by bald cypress (*Taxodium distichum*), dwarf cypress, and slash pine, the Great Dismal Swamp of Virginia and North Carolina is 80 square miles in extent and bears forest rich in Atlantic white cedar (*Chamaecyparis thyoides*) (Figure 2.4) and tupelo.

The extensively dispersed tropical wetland types are swamps and marshes. The majority of the marshland that makes up the Kakadu National Park in northern Australia is home to river red gum and coolibah trees. Swamps are abundant throughout Southeast Asia, particularly in the coastal regions. Seasonal flooding occurs in the Amazon basin of South America when the Andes snow melts in the spring. Extensive swamps are then developed over portions of the floodplain when the river overflows its banks and engulfs the surrounding forest. The vast diversity of storks, herons, and wildfowl found in Bharatpur Swamp is popular. In Africa the Okavango Swamp is seasonal, its waters extending out across the edge of the Kalahari Desert in northern Botswana. Because of the frequent floods, the river courses and water channels that pass through the wetlands are continually changing. Water naturally forms meandering channels with wide bends as it flows gently. The water may choose shorter routes during a flood, forming new channels while isolating portions of the previous channel in crescent-shaped lakes known as oxbows. In overall forest cover, we can observe open pools inside the swamp that can increase the diversity of habitats and provide new chances for aquatic plants and animals.

Figure 2.4 Atlantic white cedar

Source: https://commons.wikimedia.org/wiki/File:2013-05-10_13_29_55_Atlantic_White_Cedar_near_the_edge_of_a_bog_along_the_Mount_Misery_Trail_in_Brendan_T._Byrne_State_Forest,_New_Jersey.jpg,*Creative Commons License.*

A wetland ecosystem becomes more spatially complicated when trees are present, which allows more animals to live there. Many bird species, such as the storks, ibises, and herons that hunt in the marshes, use the trees as perches, roosts, feeding areas, and breeding places. A wider variety of fish can find refuge in the intricate underwater environment that is created by the sections of trees that are submerged. The abundance of organisms in the water and canopy serves as food for larger predators like jaguars, crocodiles, alligators, and anacondas. Thus, the swamps in the tropics are some of the most diverse and productive wetlands. Small wetlands, often measuring only a few acres, are scattered throughout remote regions of northern Russia and North America that are not near the ocean. These are the regions that the glacial ice sheets from the ice age previously covered before retreating and leaving them barren. However, the ice did not recede in a single piece towards the north; instead, it left behind islands and small areas of melting ice that resembled a landscape. The rocks and debris that the glaciers brought in covered some of these broken ice masses, so when the ice melted, it left behind a quarry full of water. The pothole mires originated from these dispersed pits. These pothole wetlands were created 10,000 years ago, but since then, soils have eroded from their margins, filling them in and making them shallow, typically less than three feet deep(Skagen et al., 2016). As the inflow continues, fens and marshes encircle them, and the surrounding vegetation serves as a shelter for ducks and other breeding wildfowl. The southern Canadian provinces of Alberta, Saskatchewan, and Manitoba, as well as the northern United States prairie region, are home to this wildfowl production. The fact that each pothole bog occupies a separate catchment and is not connected by waterways is one of their most distinguishing characteristics. This is especially noticeable from above, as the open water areas create a scattered pattern across the terrain. When the weather is rainy, particularly in the spring when the snow melts, the wetlands are filled with rainfall; but, when summer heat forces the water to evaporate, the wetlands are dryer. Clays often lie beneath the prairie pothole wetlands. Summer heat can lower water levels and sometimes even completely evaporate surface water, leaving behind muddy hollows.

Figure 2.5 Bog mosses

Source: Benavides JC, Vitt DH, Cooper DJ. The High-Elevation Peatlands of the Northern Andes, Colombia. *Plants*. 2023; 12(4):955. https://doi.org/10.3390/plants12040955,*Creative Commons License.*

Large pieces of ice that have been buried deeply in the ground can cause certain potholes to be deep. The flora that is populating the area may gradually be replaced; bog mosses (Figure 2.5) may take over if the water is acidic, and may even be able to sustain trees like white cedars.

2.7 BOGS

It is frequently the case that a small pool of open water still exists in the middle of the mire, indicating that this is how the wetland actually formed. Because the vegetation covers only a few feet thick and is spread over a body of water, the surface trembles when you walk across it. This explains why this kind of wetland is sometimes known as a quaking bog. Raised bogs can be found across North America, but they are more prevalent in the east and west of the cool temperate zone. Similar to Russia, they are widespread in northern Europe, with the coastal regions on the western and eastern sides of the continents. They can be found in New Zealand and the southernmost point of South America in the Southern Hemisphere. The unusual shape and vast extent of these bogs are most noticeable when they are without trees, as they are in oceanic regions near the sea. In more continental areas where rain and sea winds have less effect, trees cover the bog's dome and develop bog forests on top of the piles of peat. Raised bogs in Alaska are often home to lodgepole pine (*Pinus contorta*), whereas black spruce (*Picea mariana*) and tamarack (Larix laricina) are more common in the eastern raised bogs. The surface of the treeless bogs is primarily covered in bog mosses (Sphagnum), as well as a variety of sedges and dwarf shrubs, including numerous species of the heather family (Ericaceae), including bog rosemary and leatherleaf. The surface is uneven, nevertheless, with open pools of water encircled by ridges made of evergreen plants and sedges, as well as flat green patches covered with floating mosses.

Raised bogs can only form in areas with heavy precipitation and little evaporation since they are totally dependent on rain and snow for their water source(Campbell & Williamson, 1997). Because of this, they are prevalent in cold northern climates where minimal evaporation occurs. The higher precipitation that typically occurs near the seas accounts for their abundance in the more oceanic regions of continents. For raised bogs to form in the northern cold temperate zone, there must be a minimum of 19 inches of precipitation annually. In Chile, located at the southernmost point of South America, raised bogs can only be found in regions receiving at least 24 inches of rain annually; in New Zealand, which has a warmer temperature, raised bog formation requires 50 inches of rain. Because of the low precipitation caused by cold, descending air, the areas around the North and South Poles are basically deserts. The Arctic regions are rich in wetlands despite this. Because water evaporation is so slow in cold climates, soils become saturated and water builds up in pools and channels. But during the long Arctic winter, all of the water freezes, and the subsurface never thaws since the soil temperatures are so low even in the summer. In the High Arctic, the permafrost is a layer of permanent ice that presents an impervious barrier to downward water movement, and it exists below a depth of roughly 12 to 15 inches. Therefore, the water in the snow and soil either stays pools or flows over the surface in channels as the upper layers thaw in the spring. The ice that forms when winter returns and the water in the channels freezes, expands and pushes down into the ground. These kinds of ice wedges form over the terrain in complex patterns, forming elevated land encircled by drainage channels. Even the elevated regions are sufficiently moist for peat to form, though often only down to a depth of around 20 inches. These are the tundra polygon mires.

Figure 2.6 Flat sedge meadows

Source: https://commons.wikimedia.org/wiki/File:Sedge_Meadow_TGPP.JPG,*Creative Commons License.*

This intricate network of wetlands attracts a large number of migratory birds in the summer, particularly wildfowl. While the wetter areas offer plenty of food, the drier ridges and banks offer the birds secure places to nest and lay their eggs. These Arctic wetlands are home to a variety of wild species that breed there, including wild trumpeter swans and brant geese, white-front geese, and snow geese. Before winter sets in and the entire region freezes over, they all head south. Another form of Arctic wetland known as palsa mire is located to the south of these polygon wetlands, where the tundra meets the stunted trees of birch and pine. These mires are commonly found in northern Finland, as well as in Arctic Canada and Russia. Palsa mires generally resemble a patchwork of big mounds interspersed by patches of sedge and open pools. The mounds can have a circumference of up to 150 feet and are typically six to ten feet in height. Palsa development is a continuous cycle that results in the patchy appearance of the palsa mire; certain knolls are actively growing while others are deteriorating.

Flat sedge meadows (Figure 2.6) develop palsa mounds(Dredge & Dyke, 2020). The whole place freezes over in the winter, but the terrain is only partially protected by snow. The establishment of a sequence of tussocks of plants or local peat formation causes any region to become slightly higher; these areas freeze to a great depth because the tundra winds carry the snow away. These areas are the slowest to thaw in the spring and might not melt at all. Since ice expands during formation, these cold patches start to push upward as they grow, resulting in reduced snow accumulation and increased cold over the winter. Thus, the process carries on until the expanding ice core eventually takes the shape of a palsa mound. The vegetation shifts and the mound's surface get drier as it rises above its surroundings. Lichens first take over the rising mound; their pale color allows them to deflect much of the summer heat, keeping the palsa cool during the summer. Shrubs are dark in color and absorb sunlight, which causes the upper portions of the ice to melt in the summer. At that point, the thin layer of peat and plant start to disintegrate, exposing the dark peat beneath, which increases heat absorption and speeds up melting. Sedges take over the pool, and the whole thing starts over. It is clear that a usually cold temperature plays an important role in the creation of palsa since without it, the ice could not last the summer. Palsas are exclusively found in areas with an average annual temperature below 32°F (0°C) and a summer growing season of fewer than 120 days. The onset of palsa growth has clearly been impacted by climate change. Conservationists are now concerned that this unique kind of wetland could disappear due to the present climate change effects on the Arctic.

2.8 SALINE MARSHES

The salt marsh is the prevalent type of coastal wetland in the temperate zone(Pratolongo et al., 2019). These are most commonly found in the brackish conditions of estuaries, where rivers reach the ocean, or in the sheltered area behind ridges or barrier islands. They typically have large patches of mud on their seaward side, which is frequently made of organic elements the river carries down from terrestrial habitats further upstream. The river flow slows down as it approaches the approaching tide. The eroded silts, clays, and organic elements are too heavy for the slow-moving water to carry, so it deposits them as mud.

Figure 2.7 *Zostera marina* L. (Peter the Great Gulf, Primorsky Krai, Russia); **(B)**—Reproductive shoot of *Z. marina* L

Source: Razgonova MP, Tekutyeva LA, Podvolotskaya AB, Stepochkina VD, Zakharenko AM, Golokhvast K. *Zostera marina* L.: Supercritical CO_2-Extraction and Mass Spectrometric Characterization of Chemical Constituents Recovered from Seagrass. *Separations*. 2022; 9(7):182. https://doi.org/10.3390/separations9070182,*Creative Commons License.*

Plants like sea grass (Zostera), (Figure 2.7) which grows low on the shore and may spend its entire life submerged, or succulent glasswort (Salicornia) or cordgrasses (Spartina), which grow on the mud and are exposed at low tide, colonize the mud. Even the smallest particles come out of suspension and the mud builds up faster when there are plants around because they slow down the water even more. This raises the mud surface and allows more plants to grow there, creating vast meadows of plants that eventually include asters, plantains, sea lavender, and salt marsh grasses. The tide floods these plains less regularly as mud continues to build up and their surfaces continue to increase. However, less frequent sea flooding results in less mud input, which leads to the eventual development of a stable marsh that floods fewer than 100 times annually. The salt marsh vegetation divides into several zones based on how frequently they receive floodwater from the ocean each year, and each zone has its own unique community of tolerant plants. A salt marsh receives a powerful tidal flow, and the incoming water creates creeks that cut deep into the marsh. The frequency of these floods decreases as seawater overflows banks during extremely high spring tides, carrying materials onto the high marsh. As the tide recedes, water drains from the marsh, although some may still be present in pools known as pans, which grow into small ecosystems. They contain salt water from tidal flooding, and if hot, dry weather follows the flood tide, they evaporate and form saline pools. However, it is also possible that they will get a lot of rain, in which case the water may turn fresh and the salinity of the pans may drop quickly.

Only organisms with tolerance to salinity and temperature are able to survive in the pans during these periods of rapid temperature swings in the water.

Figure 2.8 Coastal wetlands

Source: https://commons.wikimedia.org/wiki/File:Coastal_wetlands.jpg,*Creative Commons License.*

2.9 COASTAL WETLANDS

Coastal wetlands (Figure 2.8) in tropical regions are more commonly covered with woody trees than with vegetation like grasses(Infante Mata et al., 2011). Only a small percentage of trees can survive in salty environments, and the mangrove swamps that make up the coastal wetlands of the tropics are primarily made up of one kind of tree. Mangrove swamps follow the same zonation pattern as salt marshes. The pioneer species that enter the deeper waters give way to less tolerant species that reside closer to the land because different tree species have varying tolerances to salinity and water depth. But in contrast to the grass forming herbs of salt marshes, the trees that are dominating offer a far more intricate structure above and below the water. Numerous fish reproduce in the water that runs between the mangrove roots, while reptiles feed on the water. The exposed mud between the mangrove roots provides another habitat where crabs and mudskippers can take advantage of the food supplies when the tide recedes. There is a distinct kind of coastal wetland found on the coastal plains of several parts of Southeast Asia, mainly on the islands of Borneo, New Guinea, and Sumatra. In these tropical locations, where there is year-round extremely high rainfall, falling leaves and tree branches do not break down and instead build up into an expanding pile of organic debris. With time, a sort of enormous compost heap forms that may be up to 50 feet deep and several miles across. It is surrounded by a dense forest with trees that can reach heights of 150 feet. These peat lands are dependent on rainfall for their water supply when they are elevated to such a height above the ground that they are no longer nourished by groundwater. For this reason, they have to be considered ombrotrophic wetlands. Despite being among the least known wetlands on the planet, these are also among the most at risk from exploitation for both peat and wood. They feature a wealth of fauna, including orangutans.

Due to their proximity to the water, coastal wetlands, both tropical and temperate, are susceptible to damage. Certain delicate ecosystems are susceptible to erosion and floods as a result of storms and tsunamis. Every wetland is changing over time. As silt comes in from eroding watersheds and organic matter is produced by the local flora, lakes and ponds are being filled with these deposits. In shallow lakes, the establishment of plants slows down water flow, which causes more sediment to accumulate. In a marsh ecology, a bed of reeds acts as a highly efficient silt trap. The presence of reeds accelerates sedimentation. As the mud surface gets closer to the water surface, other plants might colonize and the community composition shifts. This phase of development is known as succession. It occurs in a wide variety of ecosystems, including sand dunes. As conditions shift, other species adapt to the new environment to their advantage. They may even outcompete and drive out the species that unintentionally helped them establish. For example, wetland plants like willows and alders may establish a colony in a drying marsh where the sediment surface is almost at the water surface. The reeds that came before the trees are shaded by their dense canopy, and eventually the reeds die from a lack of sunshine.

There is succession in all wetlands. Because of this, no wetland is completely stable; rather, they are ever-changing. Along with the actions of plants and animals, the natural environment risks also contribute to the development of new conditions. Wetlands become unstable due to a variety of factors, including storms and floods, lightning strikes and fires, hurricanes, shifting river and stream courses, and seasonal variations. Only areas where water accumulates may support the growth of wetlands, hence the ground beneath them must be water-resistant. Water can pass through certain rocks far more readily than others. Limestone, for example, absorbs water, thus water eventually sinks through it. Additionally, water dissolves limestone, particularly acidic water, which is typically found in rainfall. Acidic drainage water creates caves, and tunnels that swiftly remove any water that is present on the surface of the rocks. Although many sand grains have been compressed and bonded together to form sandstone, there are typically still very small voids between the grains. Water can therefore seep through the sandstone and evaporate. Volcanic action creates molten rock, which then cools and crystallizes to become granite. Because of this, the rock is impermeable to water and all of its air gaps have been closed. Like sandstone, slate and shale are sedimentary rocks, having formed from particles that were deposited and progressively compressed into rock in previous wetlands or beneath the ocean. Because of the adjacent volcanic activity, shale and slate have been heated to high temperatures, in contrast to sandstone, which has not undergone any further processing after compaction. They become impenetrable to water as a result of the extreme heat closing any pores in their structure. Thus, whether or not a wetland may form at a specific location depends in large part on the physical composition of the rocks. In valleys where standing water is present, the ground is undoubtedly waterproof.

2.10 ROCKS AND FLOODS

Fine particles of material can obstruct the pores of rocks, rendering them impermeable. This may occur when eroded material from the catchment rocks is carried by the rivers pouring into a valley, settling against the porous rock surface to create a coating. Given that clay is made up of particles that are smaller than 0.00008 inches (2 μm), flowing water may easily carry it across great distances. Clay particles are embedded in a variety of rocks, including granite, and as these rocks erode in a catchment, clay is released into the drainage water(Lan et al., 2003; Niwa et al., 2016). As the water velocity lowers as it enters the valley, some of these clay particles settle and may clog holes in the underlying rocks, forming a waterproof barrier that permits the growth of a wetland. Charcoal is another tiny particle that may have a comparable role in the development of wetlands. Partially burned, and extremely resistant to deterioration, plant fragments make up charcoal. Similar to a clay layer, the particles of charcoal left behind from a wildfire are blown across the soil surface and frequently washed into valleys where they build up as a black, waterproof layer that can cause the development of wetlands. Wetlands can therefore form on the foundation provided by underlying geology, which includes the softer elements like silt, clay, and charcoal that can be deposited over the parent rocks below. However, the way the terrain is designed also affects how wetland formation occurs. Peat-forming wetlands can form even on the ridges of mountains when there is an extreme amount of rainfall. Given that water can collect in landscape hollows, wetlands tend to be more prevalent there. In areas where the underlying bedrock is water-impervious (granite, shale, or slate), wetlands are typically found in valleys or hollows in the terrain. These wetlands are referred to as topogenous, which indicates that the topography, that is, the form and contour of the landscape is what determines their existence. These are some of the most prevalent types of wetlands, particularly in areas with moderate precipitation and limited water storage capacity due to extensive catchment. The soligenous wetland is a type of wetland. This is situated in hollows, much like a topogenous wetland, but it receives water from the surface where porous, water-filled rocks contact impermeable rocks below.

At the point of intersection, the water that is falling through the porous rocks due to gravity is pushed to flow sideways across the impermeable rock surface until it emerges as a series of springs. The wetland that forms along the rock border and in the hollows below it is subsequently fed by these springs. Only in situations where two radically different types of rock collide, such as when shale is layered over limestone, can this kind of wetland be found. When this happens, the water chemistry of the resulting wetland could differ from what is initially assumed. For instance, when limestone covers shale, water containing a lot of calcium carbonate (lime) runs off the surface of the acidic rock, forming an atypically rich wetland with unique flora and fauna. The floodplain wetland is a type of wetland that depends on the overall shape of the landscape. This is found in the lowest reaches of river valleys when wide ground lies on both sides of a river that frequently meanders down a mild incline to the ocean. A river's water supply comes from a broad catchment area at this point in its growth since it has frequently traveled long distances and been joined by numerous tributaries. In the event of intense rainfall across the entire catchment, or even just a portion of it, floodwater builds up in the river channel and flows quickly downstream. The river may rupture its banks in an unexpected surge of water and spread out across the floodplain, on either side. Because the fertile soils in these floodplains, which are frequently nourished by the suspended sediments of the floodwaters, are perfect for agricultural development, human communities have frequently settled in these areas. This is how large floodplains have formed on both the Nile in Egypt and the Mississippi in North America(Syvitski et al., 2012; Pennington et al., 2017). The floods that used to engulf the terrain and sustain wetlands have been eliminated by draining the river, erecting embankments along its course, and building reservoirs to regulate water flow. Because of these actions, the floodplain wetlands lose their source of water, making them more and more rare. Since the water that flows through wetlands provides the plants with nutrients, it serves as the primary source of mineral nourishment.

Chapter 3

TUNDRAS AND THEIR DISTRIBUTION

Contents

3.1 INTRODUCTION

We can learn a lot about the tundra biome from geography, including its distribution and the factors that contribute to it. Only extremely cold climates support tundra ecosystems, which are usually found in the polar regions(Chapin III et al., 1995; Chernov, 2000). There is no tundra vegetation at either the North or South Poles. This is because the North Pole is located in the center of the Arctic Ocean, far from any landmass. The water is mostly frozen in the winter; but, in the summer, the ice becomes unstable and partially breaks. There isn't actual tundra vegetation because there isn't any land close to the North Pole. The highest peaks are devoid of any flora here as well, and the valleys at extremely high altitudes may be filled with glacier ice. The area of tundra surrounding the North Pole encompasses the Aleutian Islands, the coast of Alaska, and the northern regions of Canada. It then stretches south to include the Hudson Bay, the northern tip of Newfoundland, all of Greenland, and Iceland. Tundra in Antarctica is limited to isolated patches around the coastal regions and islands, particularly at the Antarctic Peninsula (tip), due to the dense ice cap, which is located on land below sea level. Naturally, alpine tundra is widely distributed throughout the surface wherever there are tall mountains, which can be found from the Arctic to the equator. The best places in the Americas to find alpine tundra are Alaska mountain ranges; the Rockies, Cascades, and Sierra Nevada in NW America; and the Andes chain in SW America. In Europe, the Alps and Pyrenees, as well as the mountains of Scandinavia and Scotland, are home to tundra vegetation. The huge Himalayas, the vast Tibetan plateau, and the ranges that stretch eastward through Mongolia and into northeastern Russia dominate Asia, with the Caucasus high mountains located to the west of the Caspian Sea. Africa is made up of the Atlas Mountains in the north and the Drakensberg Mountains in the south. Several extremely high mountains are located almost exactly on the equator in Ethiopia and East Africa. Other isolated mountains can be found all over the world, including in Mexico and Japan.

3.2 THE UNIQUENESS OF TUNDRA

By definition, the arctic tundra is found in the high latitudes, which are located far from the equator. Because the latitudes that encircle the planet are numbered from the equator, which has a latitude of 0 degrees, the high latitudes get their name. The North and South Poles, which have latitudes of 90°N and 90°S, respectively, are at right angles to the equator. The polar regions are known as the high latitudes and the equatorial regions as the low latitudes because the angles of the lines of latitude between the equator and the poles lie between 0° and 90°, with the greater values being closer to the poles. The majority of the biomes, including savanna grasslands, deserts, and rainforests, are found in certain zones across the globe and are often found within a specific range of latitudes. Being found in both high and low latitudes, the tundra biome is distinct from the others. Since climate ultimately directs all major biomes, it makes sense to start any investigation into tundra distribution by examining climatic causes. The most evident characteristic shared by all tundra regions, arctic and alpine, is their rough weather. Low average yearly temperatures are seen in both high latitude and high altitude, but for different causes. Generally speaking, the average yearly temperature in all tundra habitats, polar or alpine, is below 25°F (–4°C). Because it prevents tree growth or spread, cold temperatures are crucial for preserving the tundra biome. Seeds including those of birch, spruce, and pine, can germinate and establish themselves when the total temperature rises over this critical point. The margin of the polar tundra closest to the equator is typically marked by forests of coniferous and deciduous trees, while its poleward corner/edge is determined by permanent ice or the ocean. Similar to this, the forest creates a lower altitude barrier that spans the entire biome in the case of the alpine tundra(Grabherr et al., 2010). Hence, tundra exists because the temperature is too low to support the establishment of forests.

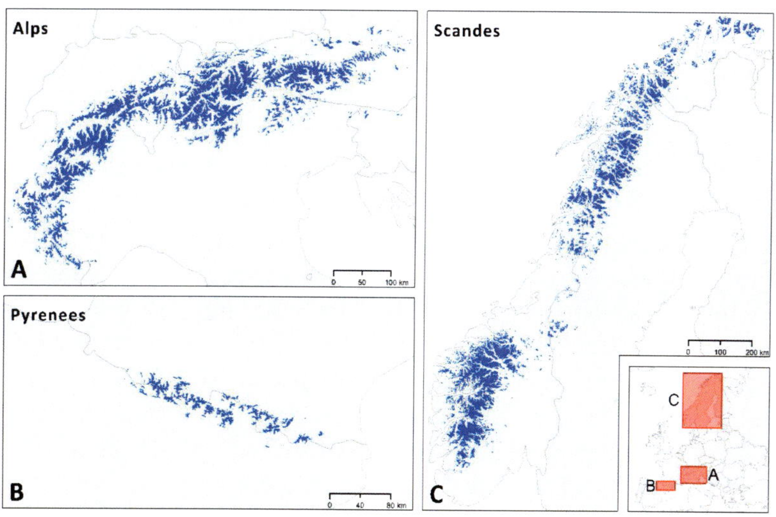

Figure 3.1 Tundra

Source: Barredo JI, Mauri A, Caudullo G. Alpine Tundra Contraction under Future Warming Scenarios in Europe. *Atmosphere*. 2020; 11(7):698. https://doi.org/10.3390/atmos11070698,*Creative Commons License.*

The yearly average temperature is not a reliable indicator of the tundra overall climate (Figure 3.1). It is important to take into account the seasonal change in temperature because a place experiencing warm summers and cold winters may have an average temperature that is similar to that of a region experiencing year-round coolness. The length of the growing season is also vital for vegetation; even in warm climates, a very short growing season may not be sufficient for the formation of forests, which could lead to the development of tundra.

However, the capacity of flora to photosynthesize and accumulate energy storage is limited, and it requires a period of time without snowfall. In addition to cold, other elements also contribute to the tundra's survival. High winds are particularly important among these. Both the high peaks and ridges of the alpine tundra and the broad stretches of the polar tundra are vulnerable to strong winds. These gusts can occasionally carry ice particles trapped in the fast-flowing air, which can be very harmful to flora if it extends far above the soil surface. Taller trees are removed, and shrubs are continually pruned using ice blasting. Both the polar and alpine tundra soils may be covered in snow or frozen for a substantial portion of the year. Tree invasion/distribution is inhibited by cold which limit the growth of roots and snow hinders the germination of seeds. Since evaporation constantly removes water from leaves, vegetation needs a steady supply of water; hence, when the weather is cold, the plant experiences stress.

This is especially true for many trees and other plants with relatively large leaves. Therefore, temperature and water are the primary climatic elements that affect plants and animals. Depending on whether they are tropical savannas or temperate prairies, grasslands can occur in a wide range of temperatures. Although deserts can exist in both hot and cold climates, they are typically found in very dry climates. Tundra has a dry and cold climate. Hence, climate is crucial to the distribution of tundra, and understanding the pattern of tundra throughout the Earth's surface requires taking into account the factors that govern global climate trends. Any land near the sea is kept warmer as a result, even in the absence of warm currents, because water loses heat more slowly than land masses do when temperatures drop. Compared to coastal areas, continental landmasses experience greater heat input and loss, making them warmer in the summer and colder in the winter. The distribution of plants may be impacted by this. In Alaska and the Yukon Territories, for instance, the border between the tundra and the forest is located at roughly 69°N, but in eastern Canada, specifically in the Hudson Bay region, the limit/boundary is located much further from the pole, at about 55°N. This is partly because of the extremely severe winters in NE America, which prevent forests from spreading further northward and favor tundra. The north of Siberia is the coldest area on Earth, and the same holds true for eastern Russia(Groisman et al., 2013).

3.3 POLAR REGIONS

It is evident that the polar regions would experience cold temperatures when all of these elements are taken into account. They receive less energy from the sun, and there is minimal effect from warm air masses and warm currents originating from the equatorial areas. The climate is warmer in areas where warm currents do enter, such as the eastern Atlantic Ocean. The polar coldest areas are those that are farthest from the ocean. The vast continent of Antarctica, which is encased in the greatest ice sheet on Earth, is located at the South Pole. The interior cannot support tundra flora because the ocean's effect is only felt at its periphery. Except for the Antarctic Peninsula, which bulges northward toward the tip of South America, almost the whole continent is contained within the Antarctic Circle. The majority of the landmass is located within latitude of 70°S. These regions, restricted to these high latitudes and surrounded by oceans, do not exhibit the Northern Hemisphere characteristic vegetation transition from polar ice to temperate woods. As a result, the Antarctic tundra is isolated on the Antarctic Peninsula, several Southern Ocean islands, and the continent's coastline border. Since the North Pole is located inside the Arctic Ocean, there is no ice sheet covering it in the Northern Hemisphere. The tundra vegetation is found on the territories that border the ocean, the North Atlantic and North Pacific regions, while the boreal woods of birch, pine, spruce, and larch encircle the tundra on its southern boundary. This boundary creates a tree line, above which only shrub growth is allowed due to stringent restrictions on the growth of tall trees. This tree line is approximately in line with the July temperature of 50°F (10°C). This average 50°F July isotherm coincides with the tree line. An isotherm is a line that connects all the locations that share a specific temperature. Healthy tree development becomes extremely challenging if the average July temperature is lower than this threshold. The tundra biome appears to be able to withstand a maximum amount of heat throughout the summer months.

The type of ground cover in the tundra habitat influences the low temperatures. Because of its consistently low temperatures, patches of snow and ice are an unavoidable aspect of the tundra landscape. Since sunlight is reflected off their white surfaces rather than being absorbed by the ground below, the presence of water in these solid forms causes additional energy loss. Albedo is the term used to describe how much a material reflects light. As a result, snow and ice have a high albedo, reflecting a large amount of light that strikes their surfaces; in contrast, many types of vegetation cover, dark rocks, and organic soils often have a lower albedo. The tundras, usually white surface contributes to the region's ability to stay cold, but the darker patches within can absorb heat and cause variations in the local climate. Instead of operating in a climate, this may do so in a microclimate, even at extremely small sizes. As the soil surface absorbs sunlight, it frequently warms up by 1°F (0.5°C) over the surrounding air temperature. For plants and other species, even minute variations can have a significant impact. The local temperature close to the soil can be raised by thin coatings of ice that are gradually melting, and by dark stones in the ground acting as heat absorbers. Only areas with an average temperature below around 25°F (–4°C) can be considered tundra ecosystems, taking into account the average annual temperature. However, average temperatures do not fully represent an area. In July, which is often the warmest month of the summer, temperatures can reach as high as 30°C on a single day, particularly in continental locations like eastern Canada or Siberia. However, because of the cool overnight temperatures, the monthly average stays below 50°F (10°C). Since they have a limited amount of time to finish their reproductive cycles, tundra plants and animals also depend on the length of the summer season. For instance, the mean temperature of Churchill, Manitoba, only rises above freezing in the months of June, July, August, and September(Gough & He, 2015). But July might be the only month in the year with an average temperature above freezing at Isachsen, in NW Canada. The species that live there are put under more stress as a result of this.

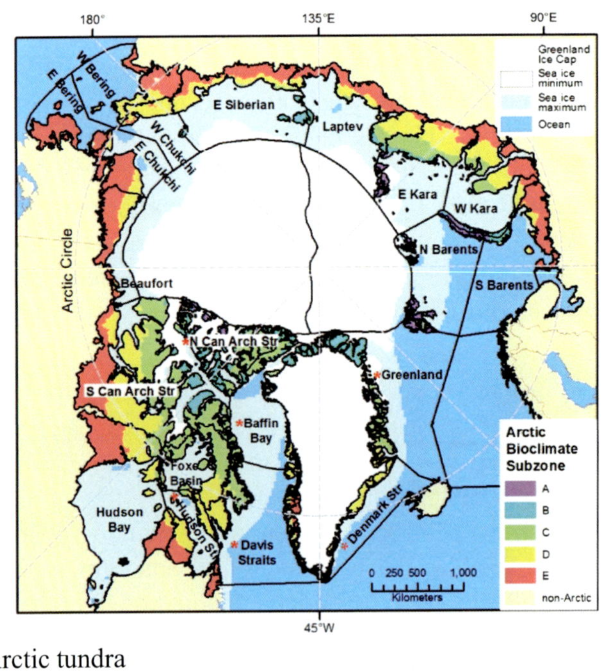

Figure 3.2 Arctic tundra

Source: Bhatt US, Walker DA, Raynolds MK, Bieniek PA, Epstein HE, Comiso JC, Pinzon JE, Tucker CJ, Polyakov IV. Recent Declines in Warming and Vegetation

Greening Trends over Pan-Arctic Tundra. *Remote Sensing*. 2013; 5(9):4229-4254. https://doi.org/10.3390/rs5094229,*Creative Commons License.*

3.4 ARCTIC TUNDRA

The Arctic tundra (Figure 3.2) climate is defined by lengthy cold winters and brief cold summers(Weller, 2000). The important seasonal variation in temperature is mostly caused by the lack of solar radiation (winter). However, because of the low sun angle and continuous night in the winter and day in the summer, the diurnal range in temperature tends to be somewhat limited. One reason the Antarctic has so few flowering plants is that it is typically colder than the Arctic, despite having large ice sheets and tall mountains. From the forest in the south to the ice in the north, the Northern Hemisphere arctic tundra is located in an area where the climate fluctuates critically. This can be conveniently divided into two zones, the Low Arctic and the High Arctic, according to geographers. The most northward groups of islands, located north of latitude 75°N, between the North Pole and Canada, Russia, Scandinavia, and Greenland are generally referred to as the High Arctic. Conversely, the southern mainland tundra regions are typically referred to as Low Arctic. In the High Arctic, the growing season typically lasts less than two and a half months, although it can span three to five months in the Low Arctic. The average July temperature in the High Arctic is only 38°F to 43°F (3°C to 6°C), whereas it is approximately 46°F to 50°F in the Low Arctic. The climatic differences between these two places have a considerable impact on the vegetation. Particularly in the northern areas, the tundra temperature is not only cold but also dry. In the northern tundra, the yearly total precipitation rarely exceeds 10 inches (25 cm). The tundra regions are hence frequently referred to as cold deserts. Anyone visiting the tundra will usually notice the low temperature immediately, although they may not notice the dryness.

Due to extremely poor evaporation, the ground is mostly wet in the summer and rich in ice and snow throughout the winter. Rainfall typically either freezes will be in the pools or travels across the surface as streams and rivers instead of evaporating back into the atmosphere. The High Arctic polar areas receive considerably less precipitation than the Low Arctic tundra ring that surrounds them. Less than 4.3 inches (11 cm) of precipitation, mostly snowfalls on polar deserts annually. This rain is like what you might get in a low-latitude hot desert. For example, the average annual rainfall in California's Death Valley is 1.5 inches (4 cm), while the Mojave Desert receives closer to eight inches (20 cm). Therefore, the name polar desert is quite acceptable given that the polar regions can actually get extremely dry. Since a multitude of factors influence the tree line and, by extension, the height at which tundra vegetation starts, the climate of the alpine tundra is more complex than that of the polar tundra. Because the temperature of the atmosphere normally drops with elevation, height itself is the defining characteristic. The lapse rate, which varies depending on a variety of meteorological conditions and geographic locations, is the rate at which temperature decreases with altitude. Because the temperature at the base of the mountains will be cooler initially, mountains in high latitudes, closer to the poles, will have tundra habitats at lower altitudes than mountains in the Tropics. It is discovered that the tree line rises toward the equator even within a single mountain range. For instance, in the Californian Sierra Nevada, the tree line is situated around 300 miles (500 km) to the north and 1,000 feet (300 m) higher at the southern end. The tree line increases by roughly 360 feet for each degree of latitude closer to the equator. Therefore, in order to support tundra, mountains in the tropics must be quite high; alternatively, they cannot support forest. The tree line rises noticeably in altitude from higher latitudes (nearer the poles) to lower latitudes (closer to the equator)(Richardson & Friedland, 2009).

3.5 ALPINE TUNDRA

Alpine tundra (Figure 3.3) has an annual average temperature that is comparable to that of polar tundra regions. Comprehensive climatic records are accessible from numerous mountain sites, including Colorado Niwot Ridge.

Figure 3.3 Alpine tundra

Source: Körner C. Concepts in Alpine Plant Ecology. *Plants*. 2023; 12(14):2666. https://doi.org/10.3390/plants12142666,*Creative Commons License.*

At an elevation of 12,300 feet (3,750 m), the average July temperature is 46°F (8°C), but the average in December and January is just 9°F (–13°C). As a result, the seasonal variance in temperature is not as great as it would be in an Arctic site like Churchill, Manitoba. This seasonal change in temperature in alpine tundra habitats is dependent on latitude. While sites closer to the equator, like the Ruwenzori Mountains of Uganda or Mount Kilimanjaro in Kenya, have little variation in their overall climate throughout the year, high-latitude mountains, those near or within the Arctic Circle, have the same wide range of seasonal variation as the Arctic tundra. On the other hand, daily temperature fluctuations in alpine tundra are frequently very large. For instance, the air temperature on Mount Kenya in East Africa, which is 13,800 feet (4,200 m) above sea level, can fluctuate between 60°F (15°C) during the day and 23°F (–5°C) at night. This can cause important issues for the plants that live there. In the arctic tundra, where the sun rarely sets during the summer and rarely rises during the polar winter, there is not this kind of difference in temperature between day and night. As a result, the difference in temperature between day and night is not very great. Wind action is another important climatic component in both the alpine and arctic tundra. Dark surfaces absorb heat, warming the layer of air closest to the ground. However, high wind speeds can disperse this warm air, particularly in areas with vegetation, where the heat is usually retained near the ground. The soil and plants become cold if this warm layer is removed. The wind is rougher when it contains ice crystals, and it is especially harmful to newly developing plants. Because of this, wind has a significant role in establishing the tree line altitude on mountains. The layout of the forest/tundra border on mountains is a good indicator of the direction of the prevailing winds. For instance, lower elevations on the western side of the Sierra Nevada mountains in California, which face the Pacific Ocean are home to tundra vegetation than on the eastern side.

3.6 THE COMPLEX MICROCLIMATE

The diverse geography of mountains adds to the complexity of their microclimate. The amount of sunlight that reaches a certain area depends on the slope and aspect. As a result, forests can reach higher altitudes on slopes facing south. Because of this, tree lines frequently have uneven extents and don't follow the natural delineations of mountains. How low down a mountain alpine tundra can develop depends on the height of the timberline. The Arctic tundra is essentially a cold desert. There is minimal precipitation brought by the air masses at the poles. Conversely, air from low-lying areas or the ocean is frequently driven upward over mountain ranges by winds at mountain sites. The water content of the air condenses as it cools, reducing its ability to hold onto water. This cloud can then precipitate as raindrops or, if the temperature is low enough, snow. The mountain side that faces the dominant winds experiences the most precipitation. The air lowers again as the intensity of the precipitation lessens in the mountain. It is reported that the lee side of the mountain experiences drier weather due to its rain shadow. Precipitation often rises with altitude. For instance, at an elevation of 5,250 feet (1,603 m) in the Rocky Mountains, west of Boulder, Colorado, the average annual temperature is 48°F (8.8°C), and the average annual precipitation is 15.5 inches (395 mm).

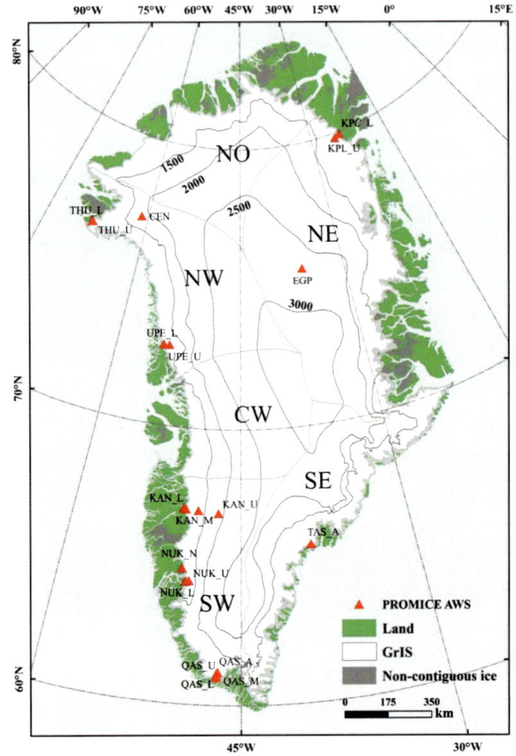

Figure 3.4 Greenland

Source: Fang Z, Wang N, Wu Y, Zhang Y. Greenland-Ice-Sheet Surface Temperature and Melt Extent from 2000 to 2020 and Implications for Mass Balance. *Remote Sensing*. 2023; 15(4):1149. https://doi.org/10.3390/rs15041149,*Creative Commons License.*

Temperature and precipitation are 42°F (5.6°C) and 21.3 inches (540 mm) at 8,460 feet (2,580 m) and 26,300 feet (3,750 m), respectively at altitudes higher than sea level. In contrast, data for Fort Yukon, Alaska, at a height of 416 feet (127 m), indicates that the average annual temperature there is 20°F (–6.7°C), with 6.8 inches of precipitation falling there each year. The shape, or topography, of the tundra varies greatly; it is not a homogenous biome. Certain areas of the Arctic have high mountains, such as Alaska Brooks Range, which constitutes the southern tundra boundary. There is a plateau and a flat coastal plain between the Brooks Range mountains and the Arctic Ocean. Although the plain contains many distinct landscape features, such as lakes, and bogs, it often gives the impression of being a large expanse of flat terrain. A large portion of the tundra of northern Canada is likewise somewhat level, and several rivers, like the Mackenzie River, flow through it(Culp et al., 2023; Emmerton et al., 2007).

In the east, where the Labrador Torngat Mountains reach elevations of almost 6,500 feet, the terrain once again becomes more rugged. There are several islands, some of the largest in the world, like Baffin Island, spread along the shoreline of north Canada.

Greenland (Figure 3.4) is a large island located further east. Its interior, which makes up 80% of its total size, is permanently covered in ice. High mountains cover its eastern coast, through which a series of mountain glaciers break out from the ice. Jan Mayen and Svalbard are two islands located in the North Atlantic. Svalbard is made up of five large islands with a coastline of deep fjords and a rough mountainous terrain. The remaining Arctic tundra is found in northern Russia and Scandinavia. The Ural Mountains, which divide the vast region in half from north to south and continue northward as the islands of Novaya Zemlya, make up the majority of this plain. The Siberian Plain is to the east of this, and the Russian Plain is to the west. These areas resemble the northern Alaskan coastal plain in many ways.

All of these areas support tundra plants and are part of the tundra biome, despite the differences in the landscape. With the exception of a few protruding mountain peaks, the majority of the Antarctic is covered by the major ice sheet in the world. As mapping tools advance, we are only now beginning to understand the terrain that lies beneath the ice. The presence of old lakes beneath the ice is a recently found feature that will surely be of interest to microbiologists, chemists, and geologists. True tundra vegetation is rare and primarily found around the coast due to the presence of ice, the extremely cold environment, and the absence of open land surface. In the few areas of the land that are free of snow and ice, patches of lichen, moss, and tiny amounts of grass can survive. The polar tundra climate varies with proximity to the poles, and this is reflected in the plant and landscape types. These tundra areas can be separated into many zones. But one must keep in mind that there are typically no clear borders; instead, the terrain progressively shifts as latitude increases. There is no zonation in the Southern Hemisphere, unlike the Arctic, because the Antarctic continent is isolated from the oceans. As a result, there is no change from tundra to forest.

Figure 3.5 Forest tundra

Source: Hauglin, M.; Næsset, E. Detection and Segmentation of Small Trees in the Forest-Tundra Ecotone Using Airborne Laser Scanning. Remote Sens. 2016, 8, 407. https://doi.org/10.3390/rs8050407,*Creative Commons License.*

3.7 FOREST TUNDRA

Forest tundra (Figure 3.5) is the term for the area in the Arctic at the northern boundary of the forest where tree development becomes challenging due to the short summer season, extreme cold, and strong winds(Epstein et al., 2004; Beamish et al., 2020). The trees are sporadic and stunted, symbolizing the southern edge of the Low Arctic. In North America, the tree line is frequently formed by white spruce on drier soils and black spruce and tamarack on moist areas. However, non-coniferous species, such as alder, paper birch, and balsam poplar, can also make up a substantial portion of the flora. Pines are abundant in Europe and Asia, especially the Scots pine, although they are rarely found in Alaska or Canadian Arctic timberline. Trees that are isolated and occasionally bent by the wind can develop in places that are conducive to their growth, or clusters of trees can live in wind-sheltered hollows. There is a thick covering of herbs and shrubs beneath the trees. This zone is a transition between the tundra and the boreal forest, also known as taiga, the biome that forms a zone in the latitudes below the tundra. Ecologists call these ecotones. The Low Arctic terrain is dominated by woody plants, which create the tall shrub tundra, outside the zone of tree development. The bushes grow to a height of less than three feet, though they can occasionally reach as high as six feet. This vegetation includes some dwarf trees and shrubs, mainly dwarf birches and willows. Other shrubs, such as Labrador tea and several cranberry and crowberry species, are included in this group. Dwarf shrub tundra takes over as the predominant vegetation type in the Low Arctic. The typical vegetation canopy is only one to two feet above the ground surface due to the loss of the higher species. In certain areas, cotton grasses may take over and the terrain transforms into what is known as tussock-heath tundra. The states of Alaska, western Canada, and eastern Siberia have notably developed tundra of this kind. The primary distinction between the vegetation types in the High and Low Arctics is that the former is primarily made up of herbaceous plants, while the latter is primarily constituted of woody plants. Any plant that extends much above ground level finds it extremely difficult to develop in this area due to the extreme cold, the short growing season, and the frequent blasting winds that carry ice crystals.

Particularly over the top of exposed ridges, where summer drought may also limit development, the wind trims the tops of these ridges and keeps them from growing higher. The local terrain affects how long the summer growing season lasts in part because protected hollows frequently have snowy patches that persist into the early summer, which limits the amount of time available for development and reproduction. In the summer, there is an abundance of water created by melting snow. Despite the slight rainfall, most low-lying areas develop wetlands because of the low rates of evaporation. Frost frequently produces intricate patterns with ridges and hollows on the ground surface. The hollows eventually become lakes with borders dominated by sedges. Wading birds and other wildfowl use these moist tundra habitats in the High Arctic as vital breeding grounds in the summer. The range of vegetation zones found in the Alpine tundra is quite similar to that found in the Arctic, however, it is graded or classified according to altitude rather than latitude. The landscape extends from the lower tree line to the exposed mountain or glacier ice. Similar to the Arctic tree line, mountain tree lines are frequently made up of stunted trees that have survived by chance in this hostile environment. The trees may grow a little higher in covered valleys than in more exposed areas, and there may occasionally be an ecotone of dwarf trees surrounding the boundaries of their growth. However, the zone is limited to the areas where the flora has naturally transitioned from alpine tundra to forest. Humans have frequently tampered with this barrier, particularly in the most densely populated alpine regions where sheep and cattle have been grazing on the high tundra and grasslands. Every summer, domestic animals are transported to the higher altitudes in many mountainous areas, where they graze on young trees as well as grasses and herbs. This extra stress has typically eliminated the krummholz zone and, in some cases, even reduced the tree line, resulting in the creation of a transition zone between grassland and tundra(Körner, 2021).

3.8 SOIL FEATURES BECOME VITAL

In certain places, soil characteristics rather than climate or human activity may determine the altitudinal limit to tree development. Grassy balds can be seen on several mountaintops in Montana and the southern Appalachian Mountains (Figure 3.6).

Figure 3.6 Southern Appalachian Mountains

Source: https://commons.wikimedia.org/wiki/File:Penns_Valley_October_2021_005.jpg,*Creative Commons License.*

Due to the shallow soils and summer drought, tree invasion is not possible in these locations; instead, a type of grassy subalpine flora grows. However, in most cases, the cold weather is what restricts the forest and allows alpine tundra flora to grow there. Because sunlight does not penetrate as well on the northern face of mountains, tree lines are frequently lower. For instance, woodland may reach up to 3,000 feet (1,000 m) on the south-facing slopes of the Brooks Range in southern Alaska, but only 2,000 feet (700 m) on the northern side. Elevated peaks in the tropics, like Mount Kenya in East Africa, exhibit a range of plant zones situated at varying elevations. These tropical plants give place to a more temperate bamboo forest at higher elevations, followed by a shrub zone of heath plants. The alpine tundra biome is located above 12,000 feet (3,600 m), and the permanent snow is found above this elevation. Similar to the Arctic tundra, distinct flora types can be seen in the alpine tundra. As observed on Mount Kenya, shrub tundra is located in lower, less exposed areas and is home to a number of plants in the heath family (Ericaceae). A few species, like bearberry and crowberry, are also found in the Arctic. In the more exposed, colder, and drier locations, grass and cushion-plant tundra can be found. It is possible to identify dry fields, rainy fields, and bare rocky areas with little to no vegetation cover, known as fellfields, within the alpine landscape. These vegetation kinds are a reflection of the growing severity of drought brought on by the topographic features like ridges and hollows, as well as the soil relative depth and drainage.

3.9 SNOW

Snow is an essential component of the alpine tundra and the Arctic. When the temperature drops below freezing and water droplets crystallize before settling, snowflakes are created in the atmosphere. Over a few days, they start to lose their crystal shape on the ground. Occasionally, they melt and then refreeze to form a more granular or powdery substance over years. We call this firn. Over the course of around ten years, the snow is completely assimilated into an ice mass, or glacier, as fresh snow falls from above and is squeezed and compacted into ice. However, not all snow is turned into glacial ice; some just melts when summer arrives. Snow patches are transient snow accumulations that might endure for a few years but never develop into actual glaciers. Every year, a few melts and leave behind a moist area of the landscape, frequently in a covered hollow with minimal sunlight and mild wind. The snow affects the local climate in a variety of ways. Because it remains in place into spring, the land is shielded from sunlight for a longer period of time. However, the snow also acts as a thermal blanket to shield the soil beneath it from the worst winter winds, keeping it warmer than the areas where the frost can get through and the snow is driven away. Varied effects of a snow patch affect both plants and animals, and as a result, distinct flora and fauna are supported in these ecosystems. Naturally, the reason compacted snow is used to build igloos is because of its ability to provide heat insulation. Because they can heat the inside of the igloo and elevate the temperature below the snow blocks much above the outside air temperature, and because the snow holds the hot air, humans are taking advantage of this insulation that snow provides. Similar to humans, many animals make use of the snow by digging burrows inside it to stay warm.

There are regional variations in the overall stability of snow. Several tundra areas, like the Siberian tundra, experience 200–280 days of snowfall annually(Eugster et al., 2000; Zeng & Jia, 2013). Although the depth varies, it is generally eight to twelve inches (20 to 30 cm) deep due to minimal precipitation. Alpine sites may have large levels of precipitation, resulting in deep snow areas. The snow depth is frequently limited by the local topography, especially slope; glaciers occur in areas where deep snow may build. In the tundra plains, local landforms can also be significant because the wind will carry snow from exposed ridges and higher areas into hollows where it will accumulate. Subsequently, under extreme heat and pressure throughout the

Proterozoic epoch, the Archaean rocks underwent changes that resulted in a compacted rock known as gneiss. Metamorphic rocks are those that have undergone these transformational processes, such as melting. These rocks most likely originally formed huge mountains, but with time, they likely eroded away, leaving behind a balanced environment. Because the rocks are usually hard and don't hold many plant nutrients, the resulting soils are acidic and deficient in nutrients. There is evidence of historical volcanic activity in some Arctic regions, such as Greenland and Jan Mayen Island. When an underwater volcano erupted in 1963, the island of Surtsey developed off the coast of Iceland. The Mid-Atlantic Ridge, a chain of volcanoes that stretches from the Atlantic into the Arctic Ocean, is linked to volcanic activity. Precambrian rocks make up the underlying geology of Antarctica. However, there are also a number of younger rocks present, especially in West Antarctica.

3.10 ANTARCTICA

Antarctica was connected to Africa, India, and Australia together with all the other major landmasses of the planet to form a single supercontinent. The climate in the past was different from that of modern Antarctica, as evidenced by fossils found in the rocks. The Earth must have been warmer during the Permian period, some 290 million years ago, as broad-leaved deciduous forests flourished near the present-day South Pole(Lee et al., 2016; Konrad et al., 2023). Growth rings found in the wood of these old trees indicate that, despite the fact that overall conditions must have been warmer than they are now, there were important seasonal variations in the environment. Geologists concluded that the continents were originally a single landmass after finding fossils of Glossopteris trees. These fossils showed that vegetation continued from Antarctica to Australia, where the fossils are also discovered. The sandstone rocks of these ranges, known as the Transantarctic Mountains, include fish fossils from the Devonian period and dinosaur fossils from the Jurassic period. The huge continent broke apart approximately 140 million years ago, during the Cretaceous epoch. Antarctica was isolated about 40 million years ago when the last land links to South America and Australia were destroyed.

3.11 GLACIERS

Mobile ice, known as glaciers, can originate in mountain valleys or radiate from ice sheets and ice caps. Compared to ice sheets and ice caps, glaciers are significantly smaller, typically about a few square miles. Glaciers can be found on high mountains closer to the equator, such as the Alps in Europe, the Himalayas in Asia, and even the mountains of East Africa, which are almost directly on the equator. Glaciers are common in the mountain regions of the far north and the far south, such as Alaska and New Zealand.

The balance between snowfall and ice melt determines the growth or shrinkage of ice sheets, ice caps, and glaciers. Since snow builds up more quickly in higher altitudes in mountain glaciers, the glacier essentially expands from the top. Since conditions are warmer downslope due to the growing weight of ice, the bottom end of the glacier will experience more melting. The rocks beneath the glacier may provide heat since the mantle, which is located beneath the crust, is hot and molten. Due to the insulating qualities of ice, which prevent heat transfer, the base of the glacier melts and becomes warmer. We refer to this process of ice loss as ablation. The equilibrium between accumulation and ablation, therefore, determines whether a given ice mass is expanding or contracting. This in turn is dependent on the topography and climate of the area. Ablation peaks in the summer and accumulation typically peaks in the winter. The water molecules would initially be buried as more snow accumulated, after which it would be compacted into the ice. After that, it would progressively recede into the glacial mass until finally emerging at the glacier base, where it would melt and disappear. Normally, ice moves in glaciers far too slowly for us to observe. The Hubbard Glacier in southern Alaska, which is one of the longest glaciers in North America, abruptly increased its pace of movement to over 33 feet (10 m) per day in 1986. High precipitation levels appear to have triggered a sudden surge in the supply of ice from one of the Hubbard glaciers, which in turn led the glacier to accelerate. Glaciers have been documented in Iceland, Greenland, Alaska, and a few places in Russia. Glacial surges have been observed in the Andes region of South America, although they are extremely uncommon in mountain glaciers outside of the Arctic.

The entire mass is pushed downward by the weight of the newly formed ice deposit at the top. This flow is further aided by the melting of the basal ice and lubrication of the glacier base caused by the warming effect of Earth's heat at the interface between the ice and the underlying rock. Lines of strain and tension are produced by the bulk movement of ice within a glacier. Deep fractures known as crevasses that run parallel to the glacier movement line may develop as a result of surface cracking(van der Veen, 1999; Jennings & Hambrey, 2021). Shear lines that occur at right angles to the direction of movement in the lower portions, can be caused by mass slippage. Melting beneath the ice can cause streams known as glacier snouts to emerge from tunnels. In these fractured areas, light may reflect a striking blue tint through the ice. This is because, even while ice reflects a large portion of light falling upon it, the majority of the light spectrum that does enter the ice is absorbed, with just blue light passing through. High ice cliffs can form when glaciers or ice sheets flow directly into the ocean; as a result, these ice walls can collapse when the glaciers or ice sheets continue to flow into the sea. It is possible for entire icebergs, or floating islands of ice, to break off and fall from the ice. Since it appears as though the glacier is producing new ones, tiny ice chunks, this process is known as calving. Glacier Bay, Alaska is now popular with tourists due to its striking ice cliffs. Because of the severe erosion that occurs in the high mountain ranges where many glaciers grow, debris, dirt, and other waste frequently end up on the glacial surfaces. As the ice scrubs the ground at its base, even more material may be eroded. All of the suspended elements are released again upon ablation, and they may be carried farther by water resulting from ice melt, a phenomenon known as fluvial outwash, or they may be deposited on the site. These deposited sediments, which are frequently located distant from any areas of active glaciers that remain, offer proof of past glaciation long after the ice has melted.

There are signs of glacier action recently in several tundra locations, both arctic and alpine. Glaciers plow land or base in a very distinctive fashion as they grind their way down a valley. A cirque, often called a corrie, is often formed when the head of the main valley is carved out in the shape of a steep, rounded basin. There are several huge boulders in Central Park, New York City, that give testament to the area's past glacial history. This shows that the tundra biome formerly covered a significantly larger area of Earth, extending as far south as present-day New York City. When glaciers recede, they leave behind elements that offer more proof of their previous size and direction. Together with massive rocks and boulders, the quantity of material carried by ice also contains finely ground debris, known as rock flour. This mixture is referred to as till or boulder clay when it is deposited in an unsorted mass. These tills, which are produced from rocks in the far north that were brought south by extensive glacier movements, cover a substantial portion of North America, northern Europe, and parts of northern Asia. There are old tills from earlier periods, but most tills are formed from glaciations that occurred during the last million years, or even less recently. These extremely old tills, often known as tillites, have hardened, cemented, and consolidated until they resemble rocks. They provide geologists hints about the timing and location of past glaciations; they are extremely valuable.

When examined, these irregularities can reveal information about the direction of a glacier flow. The orientation of the rock fragments within a till can sometimes reveal the direction of the previous ice flow. Tills made from floating ice are an exception to this rule. All of the debris that an iceberg contains is deposited into the seafloor when it melts, but the stones in this kind of sediment are not oriented. However, occasionally some minute marine fossils can be found in such material, which can aid in identifying the materials. The climate and its impact on the accumulation/ablation balance of the ice mass control the rate of glacial advances and retreats.

Figure 3.7 Glacier: **a**) The location of Waldemarbreen in north-western Spitsbergen (white arrow); (**b**) A view towards Waldemarbreen taken by UAV in August 2021; (**c**) the direction of ice flow (black arrow) at the steep glacier slope.

Source: Lamsters K, Ješkins J, Sobota I, Karušs J, Džeriņš P. Surface Characteristics, Elevation Change, and Velocity of High-Arctic Valley Glacier from Repeated High-Resolution UAV Photogrammetry. *Remote Sensing*. 2022; 14(4):1029. https://doi.org/10.3390/rs14041029,*Creative Commons License.*

A glacier (Figure 3.7) may recede after remaining stationary for a while, a few decades or perhaps centuries leaving behind a noticeable till ridge that indicates the glacier terminus. Known as a moraine, this feature frequently has a substantial impact on how the landscape will develop in the future. A lake may arise if it obstructs the flow of water from the melting glacier or the stream that comes after it. This has led to the creation of several lakes in once glaciated areas. Convict Lake is an example of a moraine glacial lake in the Californian Sierra Nevada(Clark & Gillespie, 1997; McGlue & Woolery, 2022). The sediments that build up in these lakes provide a record of the changes in the climate, vegetation, and other environmental factors that have occurred. In addition to being deposited at a glacier terminus, till may also accumulate beneath the ice, in which case the deposited materials assume a characteristic shape that is parallel to the direction of glacial flow. When the ice melts, these formations, known as drumlins, are visible. Ultimately, isolated ice chunks submerged in till gradually melt to create cavities that frequently fill with water to create kettle-hole lakes. Kettle-hole lakes have filled in to become pothole wetlands found in prairie regions of North America.

3.12 WIND, DEPOSITS AND PERMAFROST

The wind is an active agent in the process of moving glacier waste, in addition to water. The different-sized particles that come out of glaciers are carried by water, depositing them in a network of parallel, narrow ridges in the riverbed, where they eventually dry out. Strong winds that frequently occur near glaciers sweep the finer sediments and carry them over long distances before depositing them in new sites. This process occurs once the particles that make the ridges have dried. This material is continuously supplied by the rivers; thus, the wind can bring huge amounts of it and deposit it all at once, creating a silty or sandy loess soil. Thus, fine particles are abundant in loess soils. They provide fertile soil since they are made of pulverized rock, which is also rich in chemical nutrients. In particular, loess soils are common in China, Alaska, and central Europe (Belgium to Ukraine). Loess soils form the foundation of the major agricultural regions. The majority of biological material cannot survive due to the level of physical damage. Not many animals or plants often coexist in glacier settings. But occasionally, massive mammal bones and other extremely resilient objects survive glacial outwash. Since animals including humans frequently sought refuge in the caves surrounding glaciers, bones from these species may have accumulated on the cave floor, making these locations excellent places to find fossils. Over many years, the deposits accumulate in a series of strata, with the oldest being the deepest. Thus, cave deposits can reveal information on the biological history of a glaciated area over geological time. Glacial geologists have studied past glaciation events by examining these diverse glacial landforms and their traces on the terrain. Although it is reasonably simple to identify the glacial features themselves, pinpointing the precise time and order in which they were formed is far more challenging.

The term periglacial region refers to the area surrounding a glacier margin after glacial retreat(Pollard, 2018; Murton, 2021). Although the temperature is low, it is not quite cold enough to sustain an ice cover the whole year. Ice may still be present underneath the surface of the earth even after glaciers have receded, leaving behind bare rock and glacial debris. This is particularly common in arctic areas, but it also happens in plateau regions of the world like Tibet and Mongolia. Water below the surface may continue to be frozen if the average yearly temperature stays below the freezing point of water. Permanent ice is covered in moist soil because the sun warms the surface water, which stays liquid during the short summer. Permafrost is a subsurface that is frozen permanently. These impacted regions can be easily classified into two categories: discontinuous permafrost, which is made up of patches of irregular frozen soil alternating with seasonal defrosting, and continuous permafrost, which is characterized by an uninterrupted layer of permanently frozen soil beneath the earth. Permafrost becomes intermittent in the warmer, southern regions, which are typically home to coniferous forests rather than tundra. Where permafrost is discontinuous, the remaining patches are typically limited to the more exposed ridges where winter snowfall fails to cover them. Permafrost varies in depth, but it can reach very deep levels. Permafrost may reach a depth of up to two-thirds of a mile.

It is clear that the actual tundra regions are primarily found in the continuous permafrost area. The locations of permanent ice are essentially the marine complement of permafrost because the waters are always frozen. The regions with continental climates, such as Siberia, have areas of permafrost that go the farthest south. The vast area of irregular permafrost is situated to the south of the primary permafrost distribution in the Tibetan plateau. Since these areas are extremely remote from all oceans, warm ocean currents cannot affect them to get heated. Additionally, because landmasses lose heat more quickly than seas do, winter temperatures in the interior of continents are often low. The majority of the land in Tibet plateaus is higher than 12,000 feet (3,600 m), which adds to the cold temperatures(Kuhle, 1998; Aldenderfer, 2006). Despite being as far north as much of Siberia and Canada, Northern Europe is free of continuous permafrost due to the warming effect of these ocean currents. These areas are trapped in permafrost because no warm currents can reach them. Soils above the permafrost undergo a cycle of freezing and thawing that has a substantial impact on the surface. The active layer is the topmost strata or layer of soil that thaws every spring. The majority of soil chemical, biological, and physical processes occur in this layer of the soil. Only this stratum contains the roots of the plants. Plant roots only cover the top layers of the soil, or the top four inches, though they may reach down to around 10 inches. The type of soil is one of the many variables that determine the depth of the active layer. Only the top 12 to 17 inches (30 to 40 cm) of soils may melt in the summer, but in drier locations, the thaw may reach a depth of roughly three feet (1 m).

The water in the deeper soil layers that are in contact with the permafrost starts to freeze as winter approaches(Wang et al., 2009). Because the surface soil is exposed to the cold night air, its temperature drops quickly, causing the water it contains to freeze as well. The expanding soils above and below therefore put pressure on the layer of wet, unfrozen soil that is pressed between the upper and lower ice zones. As a result, a mass of thick material may be released over the nearby vegetation when bubbles of moist soil burst through the surface. When this occurs, the tundra surface becomes uneven, with patches of exposed soil that are frequently lifted above their surroundings by downward pressure. Whatever stones or pieces of rock in the soil are forced upward toward the surface by the immense pressure that the wet soil, which expands as it freezes, exerts from underneath. Frost heaving is the term for this process of pushing rocks to the surface. Stones in particular lose heat more quickly than the surrounding soil, which is why they are forced upward through the soil. In the correct conditions, stones can quickly become hot or cold due to their high heat conductivity. This characteristic of the stones causes the soil around them, especially the soil directly beneath them, to cool down more quickly, which is when water starts to freeze. The stones end up on the surface as a result of the expansion caused by this freezing pushing them upward through the soil. The frost separates the boulders into various sizes because larger stones are sometimes driven upward more forcefully than smaller stones. Furthermore, when pebbles and stones are exposed to the surface, they frequently form patterns rather than being arranged randomly. This is caused by their ability to move both sideways and upward. Stone polygons are a type of network formed by the arrangement of stones on plain ground.

3.13 UNIQUE PATTERNS

In arctic environments, stone polygons are thinner on a slope, and they produce a direct succession of stone stripes that follow the contours of steep slopes. Even in areas that are today far from the tundra, like southern Britain and New England, similar stone patterns on the surface can be found, suggesting that these areas previously experienced periglacial conditions. The creation of ice wedges in the ground is another characteristic of periglacial areas. Soil undergoes surface cracking when it freezes, and as previously mentioned, these commonly take the shape of a network of polygons. Because evaporation is minimal in the cold, water is abundant in low-lying tundra regions. Water fills these gaps, freezes, and expands in the winter, forcing the cracks deeper into the soil and causing them to widen. These features are known as ice wedges because when viewed in cross-section, the water that freezes within these fractures resembles a wedge plunged into the ground. They are smallest at their lowest tip and widest at the surface, when more water is supplied every summer. Ice-wedge polygons cover a large portion of the tundra terrain. The polygon center may sink and contain a pool of water in the summer, highlighting the pattern even more. The polygon boundary and center will both have distinct vegetation. The wedges themselves thaw out throughout the summer to create a network of interconnecting small streams. As the soils erode into the channels, these waterways fill with aquatic vegetation before becoming silty. Only extremely cold temperatures, typically with a mean annual temperature of 21°F (–6°C) or below, may cause ice-wedge polygons to form. They can be found in fossilized form in regions that are far south, which suggests that extremely low temperatures once prevailed. The polygons range in diameter from a few feet to as much as 300 feet (100 m).

When the weather is cold enough, patterns of polygons and stone stripes can also be observed in alpine environments. Polygons can be found even on the equatorial ranges in East Africa, but they are much smaller there, typically measuring only four to eight inches (10 to 20 cm) wide. They are created by the same fundamental processes as their polar counterparts, but instead of seasonal variations, the freezing and thawing are caused by variations in day and night temperatures. Maybe this explains why they are developing on such a small scale. Huge mounds, possibly reaching several hundred feet in height and diameter, can emerge from terrain in some periglacial locations. These are particularly common in the eastern Greenland coastal zone and Alaskan coastal plain. These formations, known as pingos, frequently form alongside streams and rivers or dried-out lake basins. They typically form in areas where water springs emerge under pressure from the earth and feed the ice core, which eventually freezes as it gets closer to the surface. The southern slopes of the pingo mound may have vegetation indicative of a warmer environment, potentially even a tree cover. The higher portion of the mound has a very distinct microclimate from the surrounding plains. However, if the temperature rises overall, the pingo ice core melts and the entire structure collapses to create a deep pool that is encircled by a circular rim made of soil.

In the parts of the Arctic rich in peat bogs, a similar kind of structure known as a palsa occasionally forms. Similar to pingos, palsas are big mounds created by ice developing underneath the earth's surface. Palsas are roughly 150 feet (45 m) wide and just six to ten feet tall, making them smaller than pingos in general. They are frequently found in areas of marsh where there are open pools. Small imperfections on the surface of a sedge marsh are the precursors of palsa development. An inch or two of elevation can have an impact on the amount of snow that accumulates.

The more elevated areas have less snow cover in the winter because snow is blown from the mound and collects in the hollows between. Because snow works as an insulating layer and retains some of the heat that the ground absorbs, they are colder. The hummock elevates as the ice expands during formation, causing even less snow to be trapped. The palsa surface vegetation gets drier and the drainage gets better as it gets taller. Lichens may subsequently grow to form a vegetative cover on the palsa surface. The mound stays cold throughout the warmer months because lichens that are white or light gray in color deflect sunlight. These lichens eventually pave the way to dwarf shrubs, which are darker in color and retain more heat from the sunlight. This gradually causes the surface soil to erode, exposing the peat underneath.

3.14 LANDFORMS

A variety of landforms are produced by all of these tundra-related features(Whittaker, 1993; Arendt et al., 2022). This is important when we consider the variety of microhabitats that are accessible to the flora and fauna that live in the cold tundra. There are places for mounds and hollows, marshes and drylands, slopes and cliffs, and other species to find a place to live, grow, and reproduce. The tundra biodiversity is low, but the diversity of its topography is primarily responsible for the species richness that exists there. The preservation of the biodiversity of the tundra is greatly aided by geology, or more precisely, geomorphology. As previously said, the upper layers of soil thaw in the summer, but the deeper portions of the soil remain frozen throughout the year in permafrost zones. Soils become slightly unstable as a result of the continuous heaving, and disturbance caused by the production of ice. When the wet surface soil slips over the top of the frozen lower layers on slopes, the upper layer is likely to become mobile in the summer. Solifluction is the process by which soil slides down and collects at the base. The stones are cleaved in motion by the action of mechanical damage and frost. Every soil on the planet is first created by rocks breaking down into tiny pieces. Physical and biological forces acting on the rock cause this process, known as weathering. The action of freezing and thawing is especially significant in the tundra because ice can fracture rock surfaces, causing the rock to deteriorate under the stress. This is especially noticeable in areas where temperature variations are greatest, such as cliff sides and rocks that are directly exposed to the atmosphere. On the slopes below the cliffs, freezing and thawing results in rock falls and large piles of rock debris known as scree. This is especially evident in mountainous areas, and rock deterioration may be more active in the summer at alpine locations.

The 2003 summer in the European Alps was extremely hot, causing noticeable rock falls and even forcing the closure of Matterhorn in Switzerland, a popular hiking destination, for safety reasons(Stoffel et al., 2014). Avalanches of snow frequently pose the biggest threat to people throughout the winter. They also play a significant role in the shattering and breaking of rocks. At tundra soils, annual thawing and refreezing occur in the active layer, which is where most rock weathering occurs. Such temperature variations do not happen in the permafrost, which is deeper in the soil, and the underlying rocks continue to be below freezing. As a result, physical weathering mostly takes place above ground or in the active layer. Water seeps into fractures in the rock and freezes, breaking it into smaller pieces. The chemicals found in the rocks are eventually released after being held there for many millions of years, making them once more accessible to the plant ecosystems.

3.15 WEATHERING, ORGANIC MATTER AND MINERALS

Weathering can also be the consequence of biological processes, in which living things aid in the disintegration of rocks, in addition to physical collapse. For instance, lichens that grow on the surface of exposed rocks actively attack the rock by releasing acids that erode the rock beneath them. This has the ability to remove rock particles. Moreover, algae may cover the rock surface. The rock itself may then be scraped by grazing mollusks, such as snails that consume algae and lichens. Plant roots, such as those of dwarf willows and mountain avens, may enter rock fissures and enlarge them as the roots grow. Rocks can crack easily enough by the force produced by roots spreading outward as they search for minerals and water. In the tundra, biological and chemical weathering happens more slowly than physical weathering, which happens very quickly. Due to the low temperatures, chemical weathering is slower in the tundra than in the Tropics. These chemical reactions include the impact of atmospheric acids on soil particles, which aid in the destruction of rocks. In the tundra, the process of producing acidic compounds that attack both organic and minerals is slowed down or limited to a few times of the year when the temperature is high enough to support the growth of bacteria and fungi in the soil.

The organic matter formed by the decomposing parts of plants does not decompose quickly due to the restricted activity of bacteria and fungi in the soil. As a result, soils may accumulate organic matter. This organic debris may even collect as peat in the wetter areas. Because organic matter retains water well, much like a sponge, it keeps many tundra soils from drying out too much in the summer. Additionally, peat is effective in retaining the few mineral elements needed by plants for nutrition, forming a significant reserve in the soil. The majority of Arctic rocks are comparatively deficient in the elements required by plants. There is a shortage of nitrogen and phosphorus in particular. Because of the limited precipitation, the atmosphere contributes comparatively little in the form of chemicals. The soil may grow to form a more complex profile in the Low Arctic tundra, where flora forms a continuous cover. The blueberry and heather families (Ericaceae) that make up this surface vegetation produce an extremely acidic litter when their fallen leaves fall to the ground. This dead plant litter typically builds up on the soil surface as a layer of humus, which is organic matter that decomposes slowly. A podzol is a soil that has a stratified, or layered profile. It is extensive in the taiga, or boreal woodland, although it can also be found in the southern tundra due to the presence of a dense cover of dwarf shrubs. The development of this kind of soil profile affects the vegetation and, in turn, the animals that graze on it since the A horizon of a podzol is typically acidic and low in nutrients. The production of a podzol requires a complex set of conditions. Gravity causes water from melting snow or summer rain to flow through the soil. The iron and aluminum in the soil, as well as the organic litter layer, allow the tiny clay particles to become mobile and move down the soil profile. Because of the iron oxides present, all of these moving elements end up deposited at the lower B horizon, where they create a dark reddish zone. This kind of soil profile can only arise in situations where there is relatively unrestricted water drainage through the soil since it depends on the downward movement of water.

3.16 DRAINAGE, SOIL COLOR AND PRODUCTIVITY

Poor drainage, such as low-lying sites that get water inflow from neighboring areas, frequently results in permanently wet soil below, which has a substantial impact on the chemistry and biology of the soil. Its overall appearance is also impacted by poor drainage. In such a wet Arctic location, excavating a hole would first expose a black layer of badly decomposed organic debris, and then mineral soil, often appearing blue-gray beneath it. A gley soil is one such type of soil. Given that it lacks oxygen, that is why it appears gray in color. The majority of soils have air spaces, and when air enters the soil, oxygen can get to the roots of plants, bacteria, fungi, earthworms, and other living things that need oxygen to breathe. The air cannot reach the lowest layers of the soil when it is completely saturated with water, blocking all air gaps and pathways to the top with standing water. Although oxygen may diffuse through water and dissolve in it, the pace at which it does so is approximately 10,000 times slower than it is in the air.

Thus, oxygen is frequently scarce in a water-soaked environment, depriving microorganisms, animals, and plant roots of oxygen. Chemical reactions are also impacted by oxygen scarcity. Oxygen is necessary for certain chemical processes in soils, particularly those involving the metal iron. All soils include iron, which is most commonly found in the oxidized state known as iron-III (ferric iron) in well-aerated soils. Soils typically have a deep, dark color due to the combination of iron-III and oxygen, which results in iron oxides with a reddish-rust color. But in environments lacking in oxygen, such as gley soils, iron is reduced to iron-II form (ferrous iron). The element becomes blue-gray in the iron-II state, which is clearly visible in the gley soil profile. In other words, the wet tundra soils adopt the gray color of iron found in low-oxygen environments. Upon closely examining a gley soil profile, reddish-colored specks or lines may occasionally be seen. Air can enter the soil through old root channels where the root has withered and died. They form tubes that allow oxygen to move, and when this happens, iron-III oxides, which have a distinctive rust color, occur. The way sunlight is spread in the polar tundra is unique. The days are long and the nights extremely short throughout the summer on the northern tundra. For a portion of the year, there is absolutely no darkness between the Arctic and Antarctic circles and the poles. On the other side, there is little to no daylight throughout parts of the Arctic. The day length in the alpine tundra fluctuates periodically based on its location relative to the equator(Billings & Mooney, 1968; Körner & Paulsen, 2004). There is virtually little seasonal change in day length near the equator. The quantity of photosynthesis that green plants can perform depends on the length of daylight. Since they must continue to respire throughout the dark hours, plants may need to use their stored food supplies to survive if the day duration becomes extremely short. This is because they may lose more energy via respiration than through photosynthesis. In order to minimize energy losses through respiration, many plants undergo a period of dormancy during the Arctic winter. Even on the alpine tundra, plants may choose to shut down for the winter in response to the cold temperatures and snowfall, rather than attempting to continue metabolism at a low level. In comparison to other ecosystems, the total quantity of primary production that occurs over a year is much less in the arctic ecosystem due to its short growing season. Accurately measuring primary production is difficult for several reasons. A portion of the energy that builds up takes the shape of roots, and measuring the growth of roots is difficult.

The tundra is an example of an ecosystem with low primary productivity, which typically supports a small number of animal species and has a low level of biodiversity. As soon as the light and temperature conditions in the spring become appropriate, new plant material is produced, and invertebrate grazers such as caterpillars, leafhoppers, plant-feeding beetles, weevils, and grasshoppers start to take advantage of the warm weather and the fresh food that is available. Then, as the numbers of invertebrate grazers rise, predatory invertebrates such as harvestmen, spiders, and carnivorous beetles eat them. Larger species, such as birds, also profit from the increasing amount of insects. Certain birds like the willow ptarmigan (*Lagopus lagopus*) and the rock ptarmigan live in the tundra and eat vegetation for most of the year. However, when their young hatch in the spring, they also turn to tundra insects as a source of food that is high in protein for their developing chicks. Other birds, such as wading species like sandpipers, godwits, and plovers, migrate to the Arctic tundra from their wintering grounds further south. To improve their chances of successful reproduction, they make use of the long summer days and the abundant invertebrate harvest. During this time of year in the tundra, birds that typically consume seeds, such as the horned lark, Lapland longspur, and snow bunting, also turn to insectivory to increase their protein intake before laying eggs. As the days become longer, ducks and geese move onto the Arctic tundra. Like buntings, they eat mostly grass, but they also add invertebrate proteins to their diet.

Chapter 4

DISTRIBUTION OF TROPICAL RAINFORESTS

Contents

4.1 INTRODUCTION

The Tropics are an area bordered by the equator where tropical trees are found. The Cancer and Capricorn tropics are located at latitudes 23 45°N and 23 45°S, respectively. There is one day of the year when the Sun is exactly above noon. Though they have never completely covered the tropical land area, forests can be found in any region inside the Tropics. Four regions, America and the Caribbean, Africa and eastern Madagascar, India and Malaysia, and Australia have tropical plants. Certain oceanic islands, like the Hawaiian Islands, can be included in this list. The Americas account for over 57% of the global tropical forest area, followed by Asia (25%), Africa (18%), and South America (18%) (Sloan & Sayer, 2015; Qin et al., 2017). There is terrain outside of the Tropics that is not covered in tropical trees. While there is desert to the west of the Andes in South America, montane forest develops in the mountains that extend along the western edge of Central America. Tropical and temperate grasslands replace the tropical forest of South America on the eastern side. Even though Africa is a large continent, tropical forests are only found in a limited portion of West Africa. From Guinea and Sierra Leone in the west to the Congo Basin in the east, the forest stretches along the coast. The forest is bordered by savanna or tropical grassland. There are desert regions, like the Kalahari Desert, to the south of the forest, bordered by grasslands. While grassland and desert can be found on the western side of the central mountains of Madagascar, eastern Madagascar is a shelter for tropical forests.

4.2 TROPICAL FORESTS

Tropical forests naturally cover most of tropical Asia. It is found in Sri Lanka, Myanmar, and in the western coast of India. It can also be found in Thailand and Vietnam, Malaysia, Indonesia, and Papua New Guinea, as well as the smaller islands to the east. A portion of northern and eastern Australia is also included. In most of Southeast Asia, the Amazon Basin, the Congo Basin, eastern Madagascar, and Central America have substantial rainfall. Tropical rainforests are found in these regions. The climate is seasonal elsewhere. Although there is a lot of rainfall, there are two distinct seasons linked to the summer and winter monsoons. Monsoon climates can be found in West Africa, India, Myanmar, portions of Southeast Asia, some portions of Australia, and Pacific islands. Rainfall is common in equatorial areas, but not always. For instance, the Somali Republic, which is located in the Horn of Africa, an area of northeastern Africa, is mostly desert or semi desert even though it is only a few degrees from the equator. Australia, central India, and much of Africa are too arid to sustain forests. Certain regions have mountains that trap precipitation in the air as it approaches, causing the leeward side to remain under rain shadow. Approaching air may traverse a cold ocean current, even close to the shore, where a lot of its moisture condenses, leaving the air dry by the time it reaches land. Only some regions of the Tropics support the growth of tropical forests due to climatic modifications. The majority of the Brazilian Amazon region, most of Guyana, Suriname, and French Guiana, portions of eastern Venezuela, and the western coastal districts of Ecuador and Colombia are covered by the major tropical forest (Figure 4.1).

Figure 4.1 Tropical Forest: the example of the Amazon Rainforest

Source: https://commons.wikimedia.org/wiki/File:BorneoRainforest_DSC_9267. JP Dominguez D, del Villar LdJ, Pantoja O, González-Rodríguez M. Forecasting Amazon Rain-Forest Deforestation Using a Hybrid Machine Learning Model. *Sustainability*. 2022; 14(2):691. https://doi.org/10.3390/su14020691G,*Creative Commons License.*

The forest stretches across the Brazilian Highlands in the east and the Andes in the west.

South America is home to around 60% of the global tropical rainforests. Furthermore, the Caribbean Islands and Central American tropical forests cover an area of roughly 196,000 square miles. Brazil got its name from the brazilwood tree (*Caesalpinia echinata*), which was widely distributed in the forests during Portuguese colonization in the sixteenth century. The dye, known as Brazil or Brasile, is a red pigment. The mahogany tree, or *Swietenia macrophylla*, is the tropical American tree. The heaviest and lightest timber are both produced by American trees. *Ochroma lagopus* weighs just 10 pounds per cubic foot, or 150 kg/m3. A large portion of the terrain is covered in igapós, or swamps and marshes. Trees predominate, some of which produce timber that is valued for economic purposes. For instance, *Calophyllum brasiliense* tree in South America is utilized in heavy construction. The várzea districts are flooded only once a year, away from the wetlands. Brazil nuts are still harvested from *Bertholletia excelsa* trees that grow in the forest in this area, which also produce rubber trees (*Hevea brasiliensis*). Some are small trees that grow under the huge trees in the forest, some are creeping plants that cover the ground, and several have sharp spines that keep animals away from them. Spanish moss (*Tillandia usneoides*), which belongs to the same plant family as pineapples (Bromeliaceae), is one of the many epiphytes. Similar forests can be found throughout the Central America region (Eastern). Figs (*Ficus* species) are prevalent, and lianes are abundant. The majority kind of forest in the interior of Guatemala, Honduras, and portions of Nicaragua, Costa Rica, and Panama are montane forests, which are found on higher plains or plateaus.

4.3 RAINFOREST

Lowland rain forest gives way to evergreen montane forest and subsequently a forest rich in tree ferns and mosses at higher elevations on the eastern side of Central America(Foster, 2001; Bussmann, 2002). In the highlands of southwest Costa Rica, cloud forest is prevalent above 4,900 feet and above 6,500 feet.

Figure 4.2 Oak trees

Source: Liu L, Tu Y, Li Q, Deng M. Seed Germination Characteristics of a Critically Endangered Evergreen Oak—*Quercus marlipoensis* (Fagaceae) and Their Conservation Implications. *Forests*. 2024; 15(2):235. https://doi.org/10.3390/f15020235,*Creative Commons License.*

Oak trees (*Quercus* species) (Figure 4.2) and spurges (euphorbias) can be found above the cloud forest, while evergreen shrubs, berries, heaths, and sphagnum moss can be found above 9,800 feet. The South American forest has the highest species variety in addition to being the largest tropical forest. The Amazon Basin is home to over 2,500 different types of trees. In the vicinity of Manaus, Brazil, where the Negro and Amazon Rivers converge, scientists discovered plants from 107 species across 37 families, covering an area of about 1,900 square feet. Of the 70,000 native species found in South America, about 55,000 are thought to be endemic, growing only in that region. Approximately 20,000–25,000 of the 43,000 to 48,000 species found in Central America and the Caribbean are native. The majority of the region is made up of deserts, which are surrounded by semiarid grasslands. However, almost one-third of the continent, roughly 780,000 square miles is covered by tropical forests. Part of the forest has been devastated by fires, both naturally occurring and intentionally started by locals to make way for farmland. When African forests are removed, grasslands take their place and the environment gets drier. From the Senegal River mouth, lowland rain forest forms a belt that stretches down the coast of West Africa, eventually filling the Congo Basin. This area experiences year-round heat and more than 60 inches of precipitation annually. The montane forest rises to about 4,000 feet, whereas the lowland forest stops in the mountains on the western side of the Great Lakes. Although the tropical forests of South America and Asia are more diversified than the African forest, it nevertheless contains a vast array of flora and fauna. The longest trees in the lowland reach heights of roughly 200 feet, many of which are significant for their timber. Entandophragma species yield sapele and utile, whereas certain Khaya plants yield African mahogany. The wood of *Triplochiton scleroxylon* is called obeche. Afrormosia is a timber produced by *Pericopsis elata*, and iroko is a timber produced by *Chlorophora excelsa*(Sawyer, 1992). Both are frequently used in place of teak.

Although balsa is a wood from South America, umbrella trees (*Musanga cecropioides*) in African woodlands also yield a light wood of a similar consistency. The umbrella tree is not traded commercially, although it grows swiftly and has many local uses for its wood. Diospyros species, particularly *D. mespiliformis* and *D. monbuttensis*, are the sources of ebony, a hard black wood that is also produced by African woods. Medium-sized trees called diospyros flourish in the understory. The flame of the forest, also known as the flamboyant tree (*Delonix regia*), is an attractive tree. Many forest trees have flowers. The montane woods have tree heaths above the rain forests (*Erica arborea* and *E.philippia*). There are huge lobelias (Lobelia species) and groundsels (Senecio species) that resemble trees due to their thick, unbranched stems. Massive lobelias can grow to 27 feet in height, and huge groundsels can grow to 20 feet.

The natural vegetation that covers roughly 864,000 square miles of Asia is called tropical forest. It stretches from the northern countries of Bhutan, Myanmar, and Bangladesh through portions of India, China, the Malay Peninsula, and Indonesia. Of this total area, over 566,000 square miles are lowland rain forests, and approximately 774,000 square miles are closed broad-leaved forests, or forests where the tree crowns create a continuous canopy.

Figure 4.3 Tectona grandis

Source: https://commons.wikimedia.org/wiki/File:Tree_in_new_leaves_(Tectona_grandis)_I_IMG_8133.jpg,*Creative Commons License.*

The most valuable timber, teak (*Tectona grandis*), (Figure 4.3) comes from the monsoon forests of India, Myanmar, Thailand, and Indochina. Teak is scarce in all but the most remote areas of the forest these days due to centuries of exploitation. Rain forests are not suitable for teak growth since they require a dry season. *Shorea robusta* yields wood. Its bark produces a resin that was once used to make typewriter ribbons and carbon paper, as well as a black dye. It thrives in Bangladesh and Indian monsoon forests. These days, the genus Shorea has a large number of Asia's most valuable rainforest trees. Growing in the swamp jungles of Malaya and Borneo, the *Gonystylus bancanus* tree yields ramin, a light wood used for mouldings and dowels. *Alstonia spatulata* is a tree found in the rainforests of western Malaya. It weighs only three to five pounds per cubic foot, making its wood even lighter than balsa. In addition to timber, Asian forests produce a wide range of goods. The latex used to make bubblegum is called chicle, and it comes from *Dyera costulata* trees found in Indonesia and Malaysia. Guttapercha, a sticky latex derived from *Palaquium gutta*, a rainforest tree that grows over most of the region, is used to make temporary dental fillings. When heated, gutta-percha softens but then solidifies, hardening but not breaking. It was one of the original plastics and was once used to produce golf balls and electrical insulation. Coniferous trees are uncommon in most tropical woods; Asian forests are an exception. The Chilean pine is closely related to the klinki pine and hoop pine, which are found in the jungles of New Guinea. East India copal, sometimes known as Manila copal, is a resin derived from the coniferous *Agathis dammara* tree, a relative of the Araucaria plant that thrives in central Malaysia, and is used in certain paints. Many fruits are also produced by Asian woodlands. Originally from the Pacific Islands, where it is still a staple diet in some areas, breadfruit (*Artocarpus altilis*) is now widely grown. One of the biggest fruits in the world, the jackfruit (*A. scortechnii*), is related to the breadfruit and can weigh up to 55 pounds. Mangosteen (*Garcinia mangostana*), rambutan (*Nephelium lappaceum*), and durian (*Durio zibethinus*), which is indigenous to Malaya, are other Asian fruits(Dembitsky et al., 2011; Raihana et al., 2015).

4.4 OTHER FLORA

Except for the far north, which includes Arnhem Land and the Cape York Peninsula, and the eastern coast of Queensland, where annual rainfall of at least 50 inches occurs, Australia has a largely dry climate. For instance, Cairns, Queensland, receives approximately 85 inches, whereas Darwin, Northern Territory, receives roughly 58 inches. However, because of the seasonal rainfall, this region of Australia experiences a monsoon climate with a wet summer and a dry winter. The amount of rain that falls in Cairns between the end of June and the beginning of October is a mere six inches. Darwin only gets fifteen millimeters, or 0.6 inches, during this time. November through April are the wettest months. Australian forests make up nearly 20% of the country's total land area, or over 597,000 square miles, despite the arid climate and weak soils. Australia is a country of about 3 million square miles. Eucalyptus trees predominate in the Australian forest. The genus Eucalyptus has over 450 species. Both temperate and tropical forests are dominated by them, and eucalypts typically make up the whole canopy of the forests. Acacias are commonly found in the understory

(Figure 4.4). The genus Acacia comprises about 1,200 species, of which about half are native to Australia. The majority of acacias are evergreen trees and shrubs. They are found in the drier parts of the tropical forest, which are more open and have patches of grassland and cypress pines (species of Callitris). The lowland rain forests in Papua New Guinea reach over 5,000 feet up the mountain slopes. Botanists classify it into six geographic varieties, each distinguished by distinctive trees. For instance, the Bismarck Forest system includes New Guinea walnut (*Dracocontomelum manniferum*), lancewood (*Homalium foetidum*), kamerere trees (*Eucalyptus deglupta*), and many more species.

Figure 4.4 Acacia is a valuable species

Source: Kumari M, Radha, Kumar M, Zhang B, Amarowicz R, Puri S, Pundir A, Rathour S, Kumari N, Chandran D, et al. *Acacia catechu* (L.f.) Willd.: A Review on Bioactive Compounds and Their Health Promoting Functionalities. *Plants*. 2022; 11(22):3091. https://doi.org/10.3390/plants11223091, *Creative Commons License.*

4.5 RIVERS

Many rivers are bordered by freshwater marsh forests, which are frequently covered in terentang (*Campnosperma brevipetiolatum* and *C. auriculatum*). *Metaxylon sagu*, or sago palms, are also widespread. Mangrove forests can be found growing along most of the coast; they are particularly prevalent in the Gulf of Papua, close to Madang and Lae, and in the Sepik River estuary. Along the broad, sandy beaches, forests also flourish. These include Indian beech, also known as karanja, and beefwood. Burmese rosewood, also known as Andaman redwood and amboyna wood, is used in furniture. There is a lot of bamboo, mostly on steep mountainside locations between 2,000 and 4,000 feet. Coal is used as fuel by people all around the world. Because coal is readily available and abundant. There is enough coal in North America to last for roughly 230 years if people continue to utilize it at the current rate(Pimentel et al., 1999; Jewell et al., 2019). Russia has had enough for almost 500 years, Asia and the Pacific area for roughly 145 years, and Europe for roughly 190 years. Huge reserves are also found in Antarctica, but they are hidden under the ice and cannot be mined legally. The fact that there is so much coal suggests an interesting aspect of global history. Plant organic matter that did not break down is used to make coal, swamplands are where it forms. Coal can form in swamps anywhere in the world, but it needs a moist environment, and the wettest climates are found in the tropics. Swamps are present across the planet. Many swamp forests can be found in lowland tropical forests, and in some of these regions, coal is currently developing, albeit it will take many millions of years before it is suitable for human use.

One location with significant coal reserves is Pennsylvania. This suggests that there was a tropical climate and marshy areas in Pennsylvania in the past. Large coal reserves are also found in Britain, between latitudes 50°N and 60°N. It used to have tropical weather as well. Long-term changes occur in climates. Both Pennsylvania and Britain had tropical climates and were surrounded by tropical rain forests that were bordered by deserts to the north and south when they were accumulating the plant waste that would later form coal. The South Pole was covered in an ice cap, and much of the Southern Hemisphere was covered in glaciers and ice sheets at the time 280 million years ago, although Pennsylvania and Britain were still experiencing tropical weather.

The forests were home to enormous tree ferns, 50-foot horsetails, and lycopods, which were linked to modern club mosses but could reach heights of 130 feet. Between 145 and 65 million years ago, during the Cretaceous period, flowering plants expanded widely. There were temperate forests in northern Canada, Eurasia, and Antarctica. Winter brought freezing temperatures and snowfall. There was a warmer climate later in the Cretaceous period. The climate was subtropical over most of Europe and the United States, and there was no ice at either pole.

The planet was warmer then than it is now, although some regions only experienced tropical weather since they were located in the tropics. Although forests have always grown in the tropics, tropical forests have grown in areas that are no longer in the tropics due to the ongoing rearranging or shifting of the continents. The trees that grew there were nothing like the trees of today. The transition that gave rise to modern forests took place approximately 130 million years ago.

4.6 THE UNIQUE ASPECTS OF FLOWERS

Angiosperms, or flowering plants, enjoyed an advantage over nearby nonflowering species. All of the larger plants relied on the wind to transfer pollen from male to female individuals until they appeared. They had to make enormous amounts of pollen, nearly all of which was wasted, for this to function. On the other hand, plants that flowered generated unique shoots that contained four whorls of modified leaves i.e., sepals, petals, stamens, and carpels on each shoot. The remaining organs are enclosed and shielded by the green sepals. The reproductive structures of males and females are called stamens and carpels, respectively. Collectively, they form a flower.

Figure 4.5 Conifers

Source: Cregg, B.; Ellison, D.; O'Donnell, J. Post-Emergent Control of Nuisance Cones in Fraser Fir Christmas Tree Plantations. Forests 2018, 9, 233. https://doi.org/10.3390/f9050233,*Creative Commons License.*

Many flowers have fragrant petals, and many have vivid colors. They draw insects and other species with their coloring and aroma. Pollen sticks to an animal coat when it brushes against the stamens of the flower. This pollen adheres to the carpel at its subsequent visit to a flower. Compared to only depending on the wind, this pollination technique is far more effective. Certain flowering plants, like grasses, are pollinated by the wind and have tiny and less attractive flowers. It is possible that the wind-pollinated plants never produced flowers that would draw animals, or that their ancestors did. There are further advantages that angiosperms have over their ancestors. They develop seeds that are encased in a protective structure called the ovary, and they are more effective in moving nutrients and water to their growing tips.

Conifers (Figure 4.5) and other nonflowering plants produce seeds without this protection. Eventually, a lot of angiosperms produced a fruit that was nourishing, encasing their seeds. After eating the fruit, the animals either split away the seeds or swallowed them and then expelled them. They were able to disseminate the seeds in this way than they could have if they had just fallen to the ground. When it came to self-reproduction, flowering plants outperformed all non-flowering plants. They proliferated quickly and took on a variety of shapes. Only 721 species of gymnosperms exist now, compared to around 260,000 kinds of angiosperms. Naturally, on either side of the equator contained the tropics, but similar climatic conditions might be found much further from the equator. Large tracts of tropical forests were eventually occupied by flowering plants. The majority of the region that is presently in the tropics was there 65 million years ago, but India was still in the Southern Hemisphere and was heading northward into southern Asia. The modern tropics were all covered with tropical trees. At high latitudes, the amount of water evaporating from the oceans was decreased by falling temperatures. As a result, fewer clouds formed and the tropics got drier even though they stayed warm. The rainforests could only survive close to the shore and on hilly terrain areas where annual rainfall was consistently high, while the tropical forests were pushed into a narrow equatorial strip. The subtropics gave rise to deserts, and in between the rainforests and the deserts was a seasonal tropical forest made up of plants descended from the ancestors of the rainforests.

About 20 million years ago, tropical rainforests, seasonal forests, and montane forests most likely occupied their current locations. However, this does not I mply that the tropical woods date back 20 million years. First of all, forests are dynamic and never remain the same. Because plants die and are replaced by other plants, the forest composition is always changing. Even if the forest might still exist, it might not be the same kind. The climate was still changing and the forests were forced into small spaces as the tropics grew drier. It was the beginning of the ice age. While tropical trees remained evergreen, trees in middle and high latitudes turned deciduous, losing their leaves in the fall to survive the dry winters. Over time, the global forests began to take on regional traits that finally distinguished tropical forests from temperate forests and separated forests on different continents. Nonetheless, the global environment was gradually shifting. The temperature gradient, the pace at which the temperature drops between the equator and the poles, was steepening as the higher latitudes grew colder. Antarctica was covered by ice sheets about 50 million years ago. Since most of it is above sea level, it is colder than the Arctic, which is primarily covered in water. In the Arctic, ice first formed considerably later, roughly 3 million years ago. The polar ice sheets started to move away from the Arctic as the temperatures dropped more. The planet has turned into a frigid environment where thousands of feet of thick ice occasionally cover the country as far south as the Great Lakes. The first of these ice ages, or glaciations, may have begun about 2.3 million years ago, however, the data is not very clear. Inter-glacials occurred between the start and end of glaciations. In the past 800,000 years, there have been four significant glaciations(Pages, 2016).

4.7 TROPICS

The tropics never experienced ice, and although the temperature there was still comfortable for trees, the amount of rainfall dropped significantly. Because of the reduced evaporation of water from the seas caused by low temperatures in higher latitudes, there were fewer clouds and less rain and snowfall worldwide. The ground was drier due to less rain and snowfall, which meant that there was less water on the surface of the ground to evaporate. Not just the continents but also a sizable portion of the ocean was encased in ice. When it froze, less exposed water surface area was present, which further decreased evaporation. Because tropical trees cannot withstand extended dryness, the tropical forests receded as the ice sheets moved forward. Similar to the savanna grasslands of contemporary Africa, grasslands replaced them. The forests grew again when the ice receded and the interglacial climate brought more rain, but they had to do so by dispersing their seeds from the places where they had endured the ice age. Not every plant in the forest was able to do this, and some species were extinct throughout each cycle of glaciation and interglacial. The last glacial maximum, or the coldest period of the most recent Wisconsin ice age, was roughly 20,000 years ago. The majority of scientists think that tropical forests were restricted to isolated locations, where circumstances were more manageable, during the last glacial maximum. Large swathes of the tropical forests survived, and the climate remained humid. About 12,000 years ago, as the ice age was coming to an end and the temperature and rainfall rose, the forests started to grow again. While certain areas of the current tropical forest maybe millions of years old, the majority of it is only a few centuries old and dates from the end of the last ice age. Asian forests and portions of the Congo Basin were also farmed at one point in the past. The original tropical forests are thousands of years older than the remaining intact temperate forests in the areas where they still exist.

A large portion of the most valuable tropical timber comes from trees that are dipterocarps, or members of the Dipterocarpaceae family(Rana et al., 2009). Certain trees are small, but some are huge, such as those in the genera Shorea and Dipterocarpus. These trees typically have straight, cylindrical trunks that can grow up to 98 feet and 13 feet in diameter. They also produce resins, which are used to make paints, in addition to timber. The majority of dipterocarp species are found in tropical Asia, where large forests consist of one or two species. There are fossils of extinct dipterocarp trees in Africa, where there are other dipterocarps and one species that flourish in the South American highlands. Rainfall and consistently warm temperatures are essential for the survival of many tropical species. Their seeds could not be carried thus far, much less cross the vast ocean to reach South America. They could not migrate between Africa and Asia via the Himalayas and then the Syrian, Arabian, and Sahara Deserts. This odd distribution is not exclusive to the dipterocarps. The same problem is experienced by many other plant groups, both tropical and temperate, not to mention rocks. Sandstone rocks that can only form in a hot, dry desert can be found in some areas of Devon, in the south of England. The theory of plate tectonics has replaced continental drift theory. A tectonic process is one that has to be linked with the crust deformation or the structures that result from it. The African, Eurasian, Pacific, North American, South American, Antarctic, and Australian plates are the seven main plates. The character of montane forests, which are found on the sides of tropical mountains, differs greatly from that of lowland forests. But eventually, in many millions of years, lowland forests will take the place of those highland forests, if tropical forests survive.

Additionally, some 50 million years ago, India collided with the southern corner of the Eurasian plate as it moved northward on the Indian plate. One piece of continental crust typically rests over the other as the two collapse upward. Extremely tall mountain ranges result from this. The Himalayan Mountains were raised by the collision between India and Eurasia, and this collision is still ongoing. India continues to move northward at a rate of 1-2 inches (4-5 cm) per year, and the Indian plate has been reduced in length by roughly 600 miles. Rich soils support the growth of temperate forests. Forest soils gave rise to some European farms. The soil will provide an abundance of harvests once the trees and bushes are removed and the ground is tilled. After cutting down the forest and sowing their crops, the farmers observed the crop development. After a few months, they harvested abundant crops. But after a year or two, agricultural yields started to decline. The farmers were unable to grow any crops at all unless they applied enormous amounts of animal manure and composted crop wastes as fertilizers and soil conditioners. The decline had no apparent cause, yet it persisted year after year. The agricultural area was approaching desert status. The farmers were unaware of the fact that a tropical forest differs greatly from a temperate forest because of its temperature. The same warm weather and abundant amounts of precipitation that spur rapid plant growth also create perfect growing conditions for bacteria, fungi, insects, and other animals that consume the plants.

4.8 NUTRIENTS, WATER AND FLOODS

The ground is covered in a deep layer of leaves, twigs, and other plant material in a temperate forest. This layer is thinner in the tropical forest. It is possible that the farmers saw that the roots of the trees they cut down were significantly shallower than those of temperate trees and that they frequently extended out laterally to form a dense mat that was barely below the surface of the earth. It is also possible that farmers noticed that, unlike in temperate places, animal excrement did not lay on the ground for days after it defecated. It disappeared in a matter of hours and was covered in insects in minutes. In fact, immediately upon falling to the ground, any dead plant or animal matter started to decompose. The butterflies that landed on their faces and the bees from the Halictidae family that may cover any exposed flesh were an excessive burden for the settlers to ignore. They had to have witnessed the army ant columns that ate every animal.

The decomposing remnants of plant and animal wastes seep into the soil in temperate forests(Zech et al., 1996). This continuous replenishing keeps a temperate soil fertile. In the tropics, decomposition happens too quickly. Plant roots immediately absorb the nutrients released during decomposition. Because they are getting their nutrients from below the surface, this explains why the trees have shallow roots. This implies that not much nutrient-rich soil is obtained; in fact, many tropical soils are nearly unusable. On the forest ground, it feels different from the frequent and typically strong rainfall seen at the equator. The air is so humid down there that it is hard to keep anything dry, yet the leaf canopy acts as an umbrella to protect the rain from coming down. Raindrops land on the leaves and trickle down to the lower leaves. A portion of the rain eventually drips onto the ground after trickling down tree trunks and creepers. Enough of the water evaporates during its descent to maintain constant moisture content in the air. Water that gently touches the ground might seep into the soil vertically and drain away. Rainfall causes the surface to be pulverized so forcefully that little lumps of dirt break up into tiny particles. Particles come out of the water just below the soil surface, filling in all the gaps between soil lumps to create a layer that is impenetrable to water. Rainwater then fills gaps and travels over the surface rather than draining vertically, sweeping part of the surface dirt into the closest river. When hillsides are cleared of trees, this is one of the key reasons for soil erosion. Water moves through the soil until it comes into contact with a layer of rock or hard-packed clay that is impermeable to it. Water is unable to go any lower than this impermeable layer, so it builds up above it and fills in all the microscopic gaps between the soil particles, causing it to become totally saturated. Groundwater is this type of subterranean water that can't go below a specific level; it flows downward but at a very slow pace because it is passing through soil. The water table is the upper edge of the saturated layer.

This surface is not level and smooth like the surface of a pond because it is embedded in the soil, with water flowing in both directions along the capillary fringe that separates the saturated and unsaturated layers. Due to capillary attraction, also known as capillarity, water rises through the capillary fringe and is then dragged upward through the unsaturated layer. Water molecules are drawn to solid surfaces and to other molecules, which is why capillarity arises. Water molecules are drawn to one another by their mutual attraction to form a spherical shape that permits them to be as close together as feasible. This explains why a drop of water is spherical and why its surface bulges upward when it is contained in a small tube.

Trees in tropical forests are frequently uprooted across a large region during storms. Plant debris can cover hundreds of square miles due to wind. Low pressure in the cyclone eye and onshore hurricane-force winds combine to produce storm surges, or significantly higher-than-normal tides, close to coastlines. Low-lying areas may flood as a result, and the force of the water may uproot mangroves and other coastal trees(Toorman et al., 2018). Though tropical forests have always seen tropical cyclones, the impact of these storms is prominent. Large, old trees are pulled down, but this leaves holes in the canopy that let light fall on the forest floor. Because the mature trees surrounding them provide shade from the wind, young saplings develop quickly. Simultaneously, plant matter that falls to the ground breaks down quickly, releasing nutrients that encourage the development of new plants. In a few years, the forest fills in the space. The storm may alter the relative abundance of several plant species, but the forest as a whole always recovers.

4.9 AIR CURRENTS, CYCLONES AND CHANGING CLIMATES

An atmospheric disturbance far from the sea is the precursor of a tropical cyclone. Air moves around a low-pressure area in both the northern and southern hemispheres, clockwise and counterclockwise, respectively. Clouds are created as air converges toward the center, rises and cools. It is referred to as a tropical disturbance at this point and might just last a few hours. However, if it endures longer, the winds may pick up and the pressure at its center may drop, turning it into a tropical depression. If it keeps getting stronger until the wind speed surpasses 38 mph (61 km/h), it is a tropical storm. When the sustained wind speed surpasses 74 mph (119 km/h), it changes to a tropical cyclone.

The rising air is warm and humid because the warm sea allows water to drain easily. Because of the Earth's rotation, the converging air swings to the left in the Southern Hemisphere (SH) and to the right in the Northern Hemisphere

(NH). The entire cyclone rotates due to this phenomenon, which is known as the Coriolis effect. The water vapor in the air condenses as it rises, producing heat and propelling the air upwards. A mature tropical cyclone is characterized by a spiral of clouds that can reach heights of up to 59,000 feet (18 km) when viewed from above, and within these clouds, warm air rises. There is an eye in the middle where the air is getting thin. The sky is clear, the wind is calm, and the air inside the eye is warmer than the air in the surrounding clouds. The air pressure is lowest there as well. The eye's surface air pressure is more than 10% lower than the mean sea-level pressure during the strongest storms, which can produce sustained winds of over 155 mph (250 km/h). The largest clouds are found in the eyewall, surrounding the eye. The strongest winds are located there. As one gets farther away from the eye, the air pressure increases and the wind speed decreases. Water vapor condensation provides the energy for tropical cyclones. Because of this, they require a warm sea to provide them with an abundance of moisture, and they can only grow when the sea surface temperature is at least 80°F (27°C) across a large area. A tropical forest appears as a wide, continuous carpet of dark green when viewed from above. It is simple to imagine a mass of vegetation beneath this canopy of treetops, with shrubs, climbers, seedlings, and saplings, as well as the lower parts of the full-grown trees, occupying every available space.

The vegetation modifies its characteristics in response to the changing climate. The forest bears a strong resemblance to lowland wet forests, with evergreen trees and a brief dry season. The composition starts to shift in areas where the dry season lasts longer. There is an understory formed by the smaller evergreen trees. The predominant deciduous trees that make up the upper canopy lose their leaves in the dry season to preserve moisture. Up to one-third of the trees in a semi-evergreen rainforest are deciduous, however not all of them lose their leaves at the same time(Pfadenhauer & Klötzli, 2020). More land is covered by this seasonal forest than by any other kind of tropical forest; it may be found in the center of the Amazon Basin, the wetter regions of tropical Africa, and portions of numerous Caribbean islands. It has a large number of tree species, but less than in a similar area of lowland wet forest; large stands of a single species are typical. This kind of forest is referred to as miombo in Africa, and Brachystegia, Julbernardia, or Isoberlinia species predominate over considerable portions of it. In other parts of Africa and Asia, teak in Asia, *Oldfieldia africana* in Africa, and mountain ebony make up the majority of the forest. The understory of these deciduous trees is made up of evergreens, such as the banyan tree (*Ficus benghalensis*) in Asia and eucalyptus species in some regions of New Guinea. Out of the more than 800 species of figs, the banyan is the only one whose seeds sprout from the crown of another tree, where it starts its existence as an epiphyte. Its roots grow all around the host tree, downward, and the higher portions thicken into trunks once they reach the ground and form side roots. In comparison to the trees in lowland wet forests, many deciduous and evergreen trees have buttresses, and their bark is typically thicker(Pócs, 1982). Perhaps as a response to more frequent forest fires, the bark has become thicker.

Compared to the lowland wet forest, the trees are more widely spread, allowing light to reach the forest floor. Litter dries faster as a result of this. There are places even further from the equator that receive less than one inch (25 mm) of rain each month during the two months of the dry season, between the seasonal forest and the open savanna grasslands. This environment is suitable for forest trees; however, they are mostly broad-leaved deciduous species, which are sometimes referred to as dry forests. Africa does not have dry forests; only South America and Asia do. The tallest trees are widely dispersed, and the canopy ranges in height from 60 to 80 feet (18 to 24 meters). Climbers and epiphytes are few, and while sunlight penetrates the forest floor, there aren't many herbs. Common shrubs are species of Acacia and Mimosa. Bamboos are widely distributed, and in certain areas, thickets of bamboo exist. Typically, winds are stronger near slopes. This is because wind has a chilling effect and, upon reaching a slope, it travels a shorter distance through the trees slow it down compared to the wind blowing over more level ground. Because of their high-temperature tolerance, plants thrive in tropical forests. It should come as no surprise that lowland plants cannot survive at high altitudes. Forest covers the sides of tropical mountains, although it is not the same as the forest in lowlands.

4.10 WET FORESTS

Broad-leaved evergreen plants make up most of the trees in a lowland wet forest(Tang, 2015). The height of the forest canopy is approximately 80-145 feet above ground, while developing trees can reach heights of 200-260 feet. Huge leaves and many trunks with large buttresses are characteristics of the trees. Lowland trees frequently have a trait known as cauliflory, in which the flowers are directly produced on the branches and trunks/stem. The plants that grow 1,000 feet (300 m) above sea level are typically identical to those found in the lowlands since the change in height occurs gradually. Above this altitude, however, certain lowland trees become absent, while others that are less acclimated to the lowland climate take their place. The transition to lower montane forest is more noticeable at an elevation of roughly 2,500 feet (760 m). Although the trees are not as tall and belong to a different species, they are related to those found in lowland forests. At about 50–100 feet is where the canopy is located. There are fewer trees with buttresses, and the ones that do exist are smaller. There are more climbers that cling to tree trunks. Compared to the lowland forest, there are more epiphytes, some of which are nonvascular. Over 4,000 feet, the forest characteristics drastically shift. The lowland forest trees completely vanish, to be replaced with species more common to temperate climates, such as oak and laurel. At 4,000 feet, the average temperature is 66°F, assuming that the sea level is 80°F (27°C). The higher alpine forest is found at elevations of over 5,000 feet. At this altitude, lowland trees do not grow. The majority of forest plants have tiny, needle-shaped leaves that aid in moisture conservation. This is required due to the wind drying effect. The height of the canopy is less than sixty feet. Even though they are uncommon, some trees are often a little taller than 85 feet when they do break through the canopy.

There are lots of epiphytes but not many climbers of any kind. Particularly prevalent are ferns, lichens, and nonvascular epiphytes. The forest growing on the slope at this altitude differs greatly from the forest growing in the lowlands at the mountain base. Mountain tops are frequently covered in fog. It is an extremely wet habitat that sustains unique plant groups. Because air is forced to climb as it crosses a mountain and its temperature decreases as it does so, slopes are rainy locations. As air temperature rises, the quantity of moisture it can contain increases, making rising air more likely to become saturated. Water vapor in saturated air begins to condense into liquid droplets, which form clouds. As the droplets develop and merge, the moisture within them may eventually fall as rain. When the air temperature drops below freezing on the top slopes of high mountains, water vapor condenses into ice crystals that combine to create snowflakes, and the moisture falls as snow. Clouds cover sections of the upper montane forest(Jarvis & Mulligan, 2011). This changes the forest's characteristics, turning it into a cloud forest. Sufficient moisture creates the perfect environment for the growth of ferns, which are vascular plants that reproduce by spores rather than seeds, and mosses, liverworts, and hornworts, which are nonvascular plants together referred to as bryophytes. Male bryophytes and ferns produce sperm that swim through water to identify and fertilize the eggs of female plants, as opposed to releasing pollen that can be spread by the wind or by animals. As a result, the plants can only develop in areas with a water coating that allows their sperm to pass through. The mild temperatures and wetness in tropical cloud forests result in luxurious growths of ferns and bryophytes. Because the moist air tends to be forced into these depressions between high hills and high valleys, this type of forest thrives there. The mosses in these areas, such as Sphagnum moss, which has the capacity to absorb and retain water several times its own weight, cover the ground. But mosses are not just found on the ground. Both on the ground and on tree branches, ferns can be found, with certain types of ferns on the ground being tree ferns. Bryophytes even cover the crowns of the trees.

Figure 4.6 Few Bryophyte species

Source: Kumari M, Radha, Kumar M, Zhang B, Amarowicz R, Puri S, Pundir A, Rathour S, Kumari N, Chandran D, et al. *Acacia catechu* (L.f.) Willd.: A Review on Bioactive Compounds and Their Health Promoting Functionalities. *Plants*. 2022; 11(22):3091. https://doi.org/10.3390/plants11223091, *Creative Commons License.*

Chapter **5**

TAIGA AND THEIR IMPORTANCE

Contents

5.1 INTRODUCTION

The term "taiga" (Figure 5.1) was first used to characterize the thick forests of spruce, a species of conifer tree, that are found in cold, rainy areas in northern Russia; Liang et al., 2014). The word was derived from the Latin *Boreas*, which means cold, northern coniferous forests in the 20th century.

Figure 5.1 Taiga

Source: https://commons.wikimedia.org/wiki/File:Adirondacks_in_May_2008.jpg,*Creative Commons License.*

This also includes boreal forests of North America and Europe. Taiga is another term used by some scholars to describe the coniferous forest found in mild temperate mountainous regions. Several of these forests such as those found in the Alps in Europe and the Appalachians and southern Rockies in the United States are important.

5.2 TAIGA: THE ECOSYSTEM

The taiga ecosystem has a humid climate. For every six months, there will be average daily temperatures below freezing, making the winters rough. With frequently less than 100 frost-free days during the summer, the growing season is less. Taiga differs from other lowland, coniferous forest types found in southern latitudes of the Northern Hemisphere, including the pine forests of Italy and Spain, in that it has a growing season that lasts three months or less and experiences mild temperatures. Although they are made up of entirely distinct species, conifer forests can also be found in the Southern Hemisphere, although they are also found in warmer regions. Oceans occupy the majority of the surface in the Southern Hemisphere taiga-producing areas, which range from around 45° to 70° latitude. Therefore, the Southern Hemisphere lacks a forest that is similar to a taiga. Taiga is dominated by coniferous trees. They are a part of the class Gymnospermae, which means naked seeds, which is made up of plants that bear seeds. They grow cones rather than flowers, and their leaves resemble needles. Four families comprise the majority of conifers found in taiga: firs (Abies), pines (Pinus), spruces (genus Picea), and larches, also known as tamaracks. Except for larches, all are evergreen. Being deciduous, larch trees typically lose their leaves in the winter.

Some biologists estimate that taiga makes up roughly 11% of the geographical area; this is an area that is comparable to the scorching desert and may even exceed the area covered by tropical rainforests. Because they employ different criteria to determine the specific boundaries of each biome, experts cannot agree on the actual size of the biomes. At its extremes, the taiga main lowland zone is located between latitudes of roughly 70°N and 45°N, with the majority of the biome falling between roughly 60°N and 50°N. From the western Pacific coasts of North America and Canada to the eastern Pacific coastlines of Siberia and as far east as Kamchatka and northern Japan, taiga forms a continuous belt of forest. In northwest Europe, where the North Atlantic current warms the climate and taiga is more confined. Taiga has been lost over the years in regions of Scandinavia and Scotland due to people cutting forests for livestock rearing, agricultural crops, and wood. The tundra covered in moss, lichen, heather, dwarf shrubs, and grasses, borders the taiga to the north. The growing season is less for mature trees to grow due to the limited amount of sunshine. The ecotone boundary, also known as the taiga-tundra boundary, is a zone of transition where dense, coniferous forest eventually replaces treeless tundra(Ranson et al., 2011; Guo et al., 2020). Here, patches of tundra are present. Conifers are present in the slopes where valleys stretch into the tundra. Isolated pockets of tundra-like conditions can be found in the taiga when the ground is either too cold or too wet and exposed to wind to sustain the growth of trees. Similar to its northern counterpart, the southern boundary of the taiga is an ecotone. Conifer-dominated Forest eventually changes from dense taiga to mixed conifer and broad-leaved forest, temperate grassland as it moves south. In contrast to the temperate, broad-leaved forest situated immediately south of the taiga, conifer woods have a lower level of species richness. One or two conifer species predominate over vast stretches of the taiga. For instance, the black spruce (*Picea mariana*) covers thousands of square miles of eastern Canada. The Dahurian larch (*Larix gmelinii*) is found throughout eastern Siberia, ranging from highlands to lowlands and from the northern to the southern edge of the taiga(Tchebakova et al., 2016; Kharuk et al., 2023).

5.3 BIODIVERSITY

There are two contributing factors to the low biological diversity. Only the strongest tree species survive the rough environment, especially the lengthy cold winters when there is little water available for plants to use. Since the thaw that occurred at the end of the last major ice age about 15,000 years ago, substantial areas of taiga forest have migrated northward. The taiga is not a monotonous forest cover. Even in the densely populated center of the taiga region, variations in soil drainage and climate lead to changes in the vegetation cover.

Figure 5.2 Black spruce

Source: Mallik, A.U. Post-Fire Habitat Heterogeneity Leads to Black Spruce–Kalmia L. Shrub Savannah Alternate State. Forests 2022, 13, 570. https://doi.org/10.3390/f13040570,*Creative Commons License.*

Black spruce (Figure 5.2) inhabits the hollows on north-facing hillsides in Alaska and Canada. These trees are resilient to the lack of nutrients, the saturated soil, and the cold, wet weather. The birch (*Betula*) and white spruce (*Picea glauca*) prefer south-facing slopes that are warmer, drier, and lack a deeply frozen layer (permafrost). Higher altitude, drier, warmer locations are preferred by jack pines (*Pinus banksiana*), which also grow well in areas where vegetation is frequently burned and damaged. Trees belonging to the genera Salix, Birch, and Alnus flourish in areas with moist soil near bogs, lakes, and rivers. These trees have broad leaves and are classified as angiosperms (Angiospermae), meaning they belong to the group of plants that produce seeds enclosed within a fruit or ovary.. Unlike conifers, which have needle-shaped leaves, these trees have broad, flat leaves that are with veins. Angiosperms are not like conifers in that they produce flowers as reproductive structures as opposed to cones, and their seeds are contained in fruit. In Alaska and throughout continental Canada, taiga forests predominate. The taiga covers a vast, green region of North America that stretches from Alaska in the west to Newfoundland in the east. It reaches the Great Lakes to the south and the tundra to the north. Conifers predominate here, as they do in other boreal forests. They have evolved to survive long, hard winters and nutrient-deficient soils. Open taiga with sporadic spruces, American larches (*Larix laricina*), and lichens can be found in the north. The taiga progressively transforms into a dense forest towards the south. Birches, aspens (Populus), firs (Abies), and pine trees dominate. The dense taiga can lead to an ecotone to the southeast in southern Ontario, Quebec, the Provinces of New Brunswick, and Nova Scotia. This ecotone is made up of conifers like white pine (*Pinus strobus*) combined with broad-leaved trees like American beech (*Fagus grandifolia*) and sugar maple (*Acer saccharum*) (Figure 5.3).

Figure 5.3 Sugar maple

Source: https://commons.wikimedia.org/wiki/File:Acer_saccharum_1-jgreenlee_(5098070608).jpg,*Creative Commons License.*

As we move southward from the North American lowland taiga belt, low-altitude temperatures cannot support taiga(Birks & Willis, 2008; Salzmann et al., 2008). At higher elevations, where temperatures are lower, taiga-like growth can be observed. Taiga-like trees can be found southward at higher elevations (gentle slopes). The cold climate in the northmost point of the North American Rocky Mountains, 4,900 feet, is the tree line, above which conifer forests can only grow. A mild climate surrounds the Rockies, which are located in and around Colorado, more than 1,300 miles to the south. This is where the conifer forest belt starts, at a height of roughly 6,900 feet above sea level. The tree line is approximately 9,800 feet above sea level. Dark conifers, with black spruce are found along the bog (muskeg) borders across the North American taiga. The taiga terrain is divided by lakes and rivers. Canada has many lakes, the majority of which are found in the taiga zone. The taiga of North America is home to more than 200 different kinds of birds. Approximately 200 land bird species in the United States breed in Canadian wetlands and boreal forests each summer and then migrate south to the United States for the winter. This includes finches, swifts, sparrows, and warblers. Caribou, moose, and wood bison are among the huge, plant-eating species found in the taiga, whereas brown and black bears are omnivores. The lynx and coyote are examples of big carnivores or animals that consume meat. Numerous Native Americans live in the boreal forest; individuals of Algonquian ancestry have historically resided in places east of the Great Lakes and Hudson Bay regions, as well as eastward toward the coast. Traditionally, people of Athabascan heritage have resided in Alaska, the Pacific coast, and the Rocky Mountain region. Small towns scattered throughout the taiga now rely mostly on mining and forestry, with smaller amounts coming from hunting, trapping, and tourism.

5.4 THREATS AND POLLUTION

Commercial logging for timber and wood pulp, oil and gas exploration, mining for precious minerals, and the construction of dams to produce power are the main threats to Canadian forests. Clearcutting, or the removal of all the trees in a particular area, accounts for more than three-quarters of Canadian logging activity. Such removed areas often recover slowly and are vulnerable to soil erosion unless they are planted with young trees or suitable seedlings. As the southern taiga timber supply runs out, exploitation is shifting northward into forest regions. At the moment, Canada is the main exporter of goods made of wood.

Industries in the US and Canada have been producing fumes containing sulfur and nitrogen oxides. These oxides mix with water vapor to form acid rain. Although sulfur-rich emissions from burning of coal and oil are now restricted by environmental law, many forested areas in southeast Canada are slowly recovering. Conifer leaves are damaged by acid rain. Additionally, it alters the chemistry of soil water, releasing metals that may be hazardous and reducing the amount of nutrients available to trees. Conifers are weakened by the combined impacts of acid rain, which increases their susceptibility to pests, diseases, and rough winters. The European taiga region extends to the Ural Mountains of Russia in the east and Scotland in the west. With the exception of the regions in Scotland and western Norway, the majority of the European taiga borders the arctic tundra to the north. The northernmost taiga is typically made up of patches of trees, particularly larches, scattered throughout the tundra; this is referred to as tundra forest.

Figure 5.4 Scotch pine

Source: https://commons.wikimedia.org/wiki/File:Skuleskogen_pine.jpg,*Creative Commons License.*

Scotch pine (*Pinus sylvestris*) (Figure 5.4) and Norway spruce (*Picea abies*) are found in the dense taiga of Fennoscandia(Aakala & Keto-Tokoi, 2011). After a forest is cleared for development and regrowth occurs naturally, birches and aspens usually take over. Siberian spruce (*Picea obovata*) is the predominant species in forests in western Russia. In the east, the forests are enriched with Siberian larch (*Larix sibirica*) and Siberian pine (*Pinus sibirica*). The transitional zone, also known as the southern taiga, lies south of the thick taiga and is made up of broad-leaved and mixed conifer forests. The northern limit of the English oak tree's extent generally corresponds with the southern boundary of the taiga in many places. Taiga derived land is farmed for grazing, cereals, or fodder in the southeast and combines with broad-leaved forest in the west and central regions as it extends farther south. However, taiga-like forests may emerge in their place at high altitudes. The taiga around the Scandinavia-Atlantic coast is unique. Northwest Scandinavia is extremely mild and rainy by the Gulf Stream-fed North Atlantic current, an ocean circulation. Though air temperatures hardly ever get above 59°F (15°C), the growing season lasts roughly six months. Arriving from the west, moisture-laden air cools as it is redirected up western mountain slopes, releasing its moisture in the form of precipitation. By the time this air reaches the east, less moisture is present, which results in less rainfall.

5.5 PINE WOOD

Pine woods predominate on the Baltic in Eastern Europe, growing on crystalline and sandy deposits left behind by glaciers that receded 10,000 years ago. The drier areas of the central plains are covered with spruce, fir, and, to a lesser extent, birch trees. In contrast to other regions of the taiga, the forests of eastern Europe are home to a great number of species, many of which are situated near rivers. These communities have historically been home to small-scale farmers, hunters, loggers, and occasionally herders of reindeer. After thousands of years of forest clearance or deforestation, the taiga has vanished from Scotland, with the exception of a few areas found sporadically throughout the highlands. Large taiga mammals were extinct at least several hundred years ago due to habitat degradation and poaching, including the wolf, brown bear, beaver, and lynx. The pine forests were replaced by moors of heather, and the regeneration of forest was hindered by sheep and red deer grazing. The last 200 years have seen the exploitation of the taiga in much of western and central Finland. Here, the focus has been on logging for timber, with clear-cutting portions of the forest and the planting of trees of the same age and species to replace them before they are also harvested. The biological richness of the forest is significantly diminished by such actions. For instance, 1,900 woodland species in Sweden are listed as endangered by the IUCN on their Red List of plants and animals(Lundell et al., 2015). A diverse forest system is currently being raised for diverse purposes, including timber production and outdoor activities like boating, bird watching, and hunting. Although the taiga has been inhabited in European Russia for thousands of years, humans have not completely destroyed the forest trees, unlike what has happened further west. Nonetheless, the forest has been greatly impacted by human activity. For agriculture, people have lit fires in the forest, usually to make hay fields for feeding cattle and other animals. In certain places, the forest has returned, although the variety of tree species has changed. The animal population has changed as a result of hunting and fishing. For the past century, most of the taiga has suffered with selective logging of high-value trees by foresters. The Siberian taiga reaches a maximum width of roughly 1,000 miles from north to south as it stretches from the Verkhoyansk Mountains in the east to the Ural Mountains in the west. It is the largest taiga area in the world.

The West Siberian plains have hot summers and extremely cold winters as the Ural Mountains sustain rain clouds from falling on the region. This is the best place for Siberian pine and spruce, although broad-leaved trees may not grow easily in winter. Except for the southern transition zone where birch and aspen grow between the taiga and the Russian steppe, they are barely noticeable. Anticyclones, which are high-pressure weather systems that transport cold air masses from the Arctic, have a significant impact on the Siberian climate. The cold, rainy weather in western Siberia promotes the formation of bogs on a large scale. Only some trees can withstand the short growth season brought on by the combination of continental dryness in the Tien Shan Mountains (Kazakhstan), located further south(Zubairov et al., 2019). Between the Yenisey and Lena Rivers, central and eastern Siberia has a continental climate, with summer temperatures that are typically higher than those in western Siberia and less annual precipitation in the form of snow, rain, and hail. The permafrost is thicker in areas where there is less snow on the ground, depriving the trees of usable water during the winter. Much of north and central Siberia is home to the resilient Dahurian larch, but only in the south, where temperature fluctuations are less severe, do spruces, pines, and firs become abundant. Birches quickly take over areas where conifer tree cover has been lost due to fire or logging. With scorching summers and icy winters, the climate is more continental east of the Lena River and away from the Pacific Ocean. In Verkhoyansk, the average monthly temperature difference from January to July is more than 108°F (60°C). In portions of eastern Siberia, the summer drought promotes grasses over forests. Large stretches of steppe covered in grass can be found here, interspersed with forests. The Siberian stone pine (*Pinus sibirica*) is the most common tree in the mountains. There are far fewer bogs in eastern Siberia than in the west due to the rugged terrain and lack of precipitation. The Siberian taiga, which is even bigger than the Amazon rainforest, is the largest region of old-growth forest in the world. In Russia, the largest populations of large taiga mammals, such as reindeer (caribou), elk (moose), wolves, red foxes, and brown bears, are found in central and eastern Siberia. Many larger animal species are kept from migrating because of the Yenisey River, which serves as a barrier. At least 11 different species of birds and mammals can be found on the east side of the Yenisey but not on the west, and reindeer and moose are more abundant in the east than in the west.

5.6 MORE DETAILS

The southern Siberian taiga is being damaged by logging, and oil and gas exploration, and these activities are slowly advancing north(Bergen et al., 2013). Large areas of taiga forest have already been damaged by air pollution from metal-smelting factories, and rivers have been contaminated by the metal-rich wastewater discharges. A number of proposed hydropower projects pose a risk to sink portions of the southern taiga.

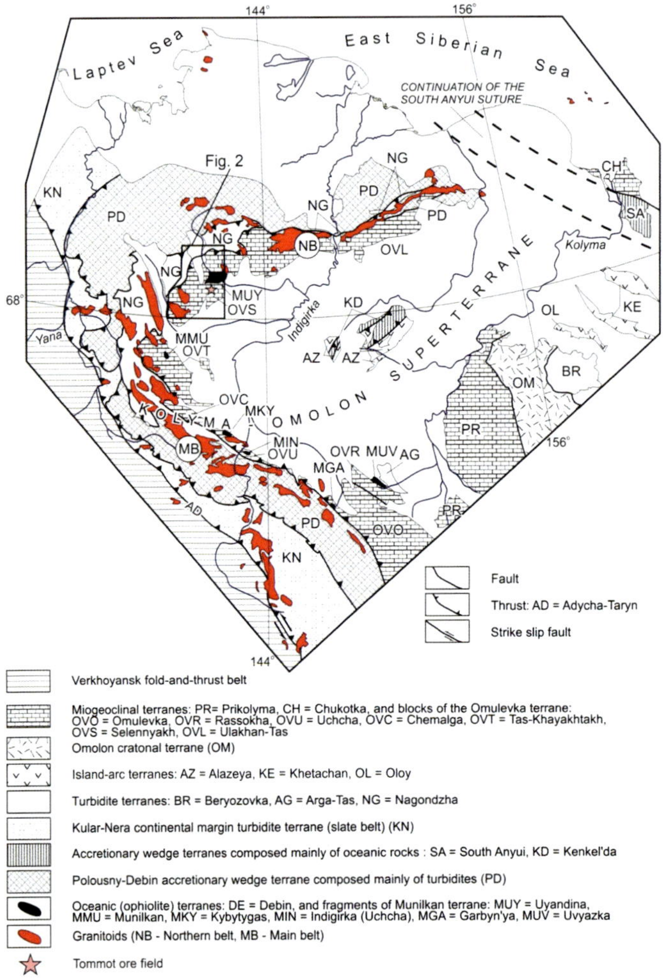

Figure 5.5 Verkhoyansk Mountains

Source: Trunilina VA, Prokopiev AV. Petrology of Granites of the Tommot Rare-Earth Ore Field (Verkhoyansk–Kolyma Orogenic Belt). *Minerals*. 2022; 12(11):1347. https://doi.org/10.3390/min12111347,*Creative Commons License.*

The Verkhoyansk Mountains, (Figure 5.5) which border eastern Siberia, form the eastern edge of the East Asian taiga, which continues eastward to the Sea of Okhotsk, where it is fragmented up by tundra, and ends at the remote Kamchatka peninsula. Taiga reaches the Korean peninsula and the mountainous areas of northeastern China to the south. Dense taiga forest extends as far south as 43°N on the northern Japanese island of Hokkaido. The Kamchatka peninsula has volcanoes with streams, boggy tundra, and taiga areas. With an area nearly equal to that of Japan and a length of 745 miles, Kamchatka is home to over 300,000 people. Kamchatka is cooled by the Oyashio current, which flows from the north. Sea mists are produced as this current collides with the warm Kuroshio current, also known as the Japan Current, which is flowing from the south. For nine months of the year, there are intermittent spells of mist and drizzle that only clear during the hottest summer months. Taiga has a unique combination of geological and climatic features, as well as a low human population. Volcanoes provide heat and nutrients to the soil, which is beneficial to plants. In the lowlands, spruce and larch flourish. Along the coast, stands of stone birch and willow have been twisted by cold winds. Only in Kamchatka cana we find the native silver fir species (*Abies gracilis*), a remnant of preglacial times over 100,000 years ago. It is currently limited to a small area of forest that is less than 50 acres in size. When salmon arrive to spawn, the rivers, which are unobstructed by dams or other man-made barriers, fill up with a large number of fish. Different varieties of salmon, sockeye, pink, chum, and silver ascent Kamchatka rivers at different periods offer a feast for the brown bears. The coastal areas of eastern Siberia and the region known as Ussuriland, which stretches from the Amur Estuary in the north to the borders of China and Korea in the south, have different climates. While the Siberian anticyclone prevails in the winter, depriving this region of rain and bringing cold temperatures, the summer monsoon winds of East Asia bring moderate weather and abundant amounts of rain from the Pacific Ocean. Taiga tends to thrive at latitudes where deserts are prevalent in central Asia to the west due to the combination of cold winters and abundant summer moisture.

The taiga in the northern regions of Ussuriland is dominated by spruce, birch, and stone pine. Stands of Korean pines (*Pinus koraiensis*) are scattered throughout, and their enormous cones are filled with up to 200 pine nuts, or bare seeds, which are highly oil-rich. These nuts provide food for a wide variety of species, including mammals like deer and wild boars as well as birds like woodpeckers. Top predators include wolverines, lynx, and brown bears. The Khingan fir (*Abies nephrolepis*) and Yezo spruce (*Picea jezoensis*) are the dominant trees where the monsoon winds have the main impact. The taiga vegetation in the southern transitional zone along the Russian southeast border with China is remarkably diverse, with several kinds of trees, shrubs, and climbers flourishing in the dark, and acidic soils. Some of the rarest terrestrial mammals can be found in these forests.

5.7 PODZOL

Taiga soils are mostly podzols. The texture and gray color of the soil just below the surface are best described as ash-colored. Podzol is known as spodosol in the U.S. soil classification system. Podzols can form in the taiga's cold and moist climate. The bacteria that decompose pine needles and other plant and animal waste that end up on the forest floor are inactivated by cool temperatures. Therefore, there is a layer rich in humus, partially decomposed animal and plant matter, just below the surface of the earth. Precipitation also exceeds evaporation. Throughout the year, there is a larger amount of water that falls to the ground than evaporation from the ground surface. Water moves through the soil generally downward under such conditions. Plant nutrients that have dissolved, including phosphates and nitrates, are carried by the running water. Metals are captured by plant decomposed products, which then transport them deeper into the earth. The near-surface soil loses nutrients as a result of leaching, the general process by which soluble substances are transported away from the upper soil by percolating water. With pH values normally in the 3.5–4.5 range, the soil is also acidic. The primary cause of this is that silica, the primary component of sand, is usually abundant in the rock that creates taiga. Acid soils are typically produced by silica-rich particles. The acidity of the soil is increased by the remnants of conifer trees, which fungi break down to release acids. Conifers grow better in these nutrient-poor, extremely acidic conditions than most broad-leaved plants(Zhang et al., 2023).

Typically, a podzol has a gray layer rich in clay that extends a foot or so below the dark brown, humus-rich top layer, from which most of the metals and plant nutrients have been washed out. Usually, a reddish-brown layer rich in iron and aluminum that was deposited by the descending acidic water lies beneath the gray layer, which is often followed by a thin layer of nutrient-rich material that was leached out of the humus. The metal-rich layer can get so compressed that very little water can get through. If so, the soil above the layer might get wet, which would be perfect for the development of peat. The reddish-brown layer rich in metals usually is present above bedrock or damaged glacial deposits. Taiga plants primarily obtain their nutrients from the upper layers of the podzol, which are rich in decomposing leaves, and the lower layer, which is rich in nutrients; roots naturally grow toward these areas. Numerous conifers yield tiny seeds, such as those of fir, pine, and larches, which are poorly suited to germinate in the leaf litter found on the forest floor. The deteriorating tree trunks in the forest are the ideal growing medium for young plants. Bogs are low-lying places where water covers the ground for most of the year. Oxygen from the air enters bogs more slowly due to the high-water content. Consequently, this attenuates the activity of microorganisms responsible for decomposing plants. This gradual breakdown, when combined with the low temperatures, promotes the formation of peat, a partially degraded layer of dark brown plant debris. Peat accumulations have the potential to be compacted and buried underground for millions of years, giving rise to coal reserves. Dried peat itself is a low-grade fuel. In addition to plants, animal bones, and human artifacts also take a long time to decay in peat. Peat deposits contain a lot of nutrients that are not easily recycled. Plants growing in peat bogs frequently have to suffer from both a shortage of oxygen and nutrients(Laiho, 2006).

5.8 ADAPTING TO LIFE IN A TAIGA

Sphagnum mosses inhabit the lake border because they can withstand the limited nutrient supply and low oxygen levels. The area of open water decreases as the sphagnum mosses move toward the lake center. Sphagnum grows upward toward the light, blocking out sunlight for the moss layers below. The moss beneath, deprived of light, dies and becomes part of the peat. The peat layer increases, gradually transforming the lake borders into a bog that climbs higher and starts to dry out. Open water has been replaced by forest. A few feet below the surface, much of the northern and eastern taiga in both Eurasia and North America is permanently frozen. This frozen layer, known as permafrost, is found in many areas of eastern Siberia, northern Canada, and parts of Alaska. It can reach thicknesses of over 1,000 feet. The heat from the interior keeps the ground from freezing below the permafrost. In about one-third of the taiga region, permafrost has a significant impact on the vegetation. It tends to make soils soggy, inhibits the downward movement of water, and promotes rainwater to run off the land. Frost heave is caused by substantial pressures in the ground caused by the top layer of permafrost thawing in the summer and then refreezing in the winter from the top down. When cold air temperatures seep into the soil, blocks of ice accumulate close to the soil surface, causing frost heave. Since water expands when it freezes, ice that is building in excessively damp soil attracts more water, which eventually freezes and causes the ice block to develop. As a result, the soil becomes filled with ice, pushing the soil above it higher. Frost heaving causes the soil to become partially mixed as the ice block progressively melts. This mixing causes the soil's deeper layers to rise to the surface in patches. Frost heave can mix the soil to depths of about 12 inches when it occurs year after year. Depending on the time of year and the equilibrium between thawing and freezing in the uppermost layers of the soil, water in permafrost-affected soils may rise or sink.

Projects involving ground excavation, like building construction or oil prospecting, are complicated by permafrost. When permafrost is exposed to warm air through digging, the top layer melts. The water remains in the soil because it cannot escape through the frozen soil below. When soil gets wet, it acts like a slowly flowing liquid, which can be problematic for anyone installing pipes, buildings, industries, or roadways. People in colder regions of Siberia have been building their homes on supports for decades to prevent the permafrost from thawing due to anthropogenic heat. In order to prevent the issue of permafrost melting, engineers developing the Alaska pipeline, which transports oil southward into the United States, also chose to construct lengthy sections above ground. Similar above-ground pipelines were built by Siberians, but because of inadequate ground insulation, the pipelines cracked, resulting in oil contamination accidents. The terrain that was once extracted by advancing glaciers and covered with deposits left behind by retreating ice is now home to the taiga biome. This geological process has occurred multiple times in the previous two to three million years. About 20,000 years ago, the most recent glaciation, a time when glaciers and ice sheets were extensive in northern Eurasia and North America peaked(Stuart, 1991; Rapp, 2019). As the ice has receded over the past 11,000 years, the gaps left behind have accumulated precipitation to create many lakes. Taiga soils are young, having formed within the last 10,000 years on deposits or bedrock. Uneven soil conditions are caused by variations in the underlying material and how it drains. Certain locations are damp, others arid, and some nutrient-rich. The variety of plant communities that may be found in the taiga today is a result of this. Taiga grows in areas with humid climates and cold temperatures for most of the year.

The majority of taiga trees have exceptional resistance to cold, but they can also withstand heat as long as there is humidity present. Eastern Siberia experiences an incredible 180°F (100°C) temperature variation throughout the year, with winter temperatures as low as -76°F (-60°C) and peak temperatures of -104°F (-40°C). Even in such conditions, the Dahurian larch manages to survive. For taiga trees, summer moisture in the air is essential. Without it, even on damp ground, they cannot develop to a large size because they lose water via their needle-like leaves too quickly. Under typical taiga conditions, these trees would grow to heights of over 100 feet. Scientists assess a variety of technical aspects of weather, including air pressure, humidity, and evaporation. The average of the weather patterns from year to year and decade to decade is referred to as the climate. It is the extended pattern of a region's temperature, air pressure, precipitation, cloud cover, wind speed, and other factors. People often discuss the weather in the United States and Canada's Great Lakes region. Conversely, the general patterns of weather observed worldwide are referred to as the global climate. The layer of air that envelops the surface is known as the atmosphere. At the equator, it is roughly 560 miles (900 km) high, whereas at the poles, it is slightly lower. The oxygen that most plants, animals, and microscopic organisms require for respiration, the process of breaking down food materials to generate energy as well as for human breathing is found in the atmosphere. Water circulating from the sea is fed into the land by its clouds. It serves as an insulating layer, preventing the Earth from warming or cooling down too quickly for life to exist on it. The air that circulates in the troposphere, the lowermost layer of Earth's atmosphere, determines the weather and climate. The environment serves as a barrier as well. Oxygen is found in the stratosphere, an intermediate layer of the atmosphere that absorbs ultraviolet (UV) radiation from sunlight. The stratosphere ozone layer is an indicator that dangerous UV radiation has been absorbed. Elevated UV exposure can cause mutations, which are alterations in a cell's genetic composition that might result in cancer.

5.9 CLIMATE AND THE UNIQUE WORLD OF LICHENS AND MOSSES

Global weather and climate patterns are influenced by solar heating, which is the uneven heating of the Earth's surface by the Sun. The area of the Earth where the Sun shines directly overhead for at least one day per year is located between the tropics of Cancer in the Northern Hemisphere (23.5°N) and Capricorn in the Southern Hemisphere (23.5°S). The area is referred to as the Tropics. Because it receives more sunlight overall, both in terms of intensity and duration than the poles. Tropical air rises as it warms. The rising air is replaced by low-level cool air that is moving in from higher latitudes. The warm air rises until it reaches the cool tropopause, which separates the troposphere from the stratosphere. After that, the air moves toward the poles through the upper troposphere. The ascending air expands as the air pressure decreases. The air molecules disperse more. Adiabatic cooling is the result of fewer collisions producing a cooling effect that lowers the air temperature. Because the upper troposphere transfers heat into space and via the upper atmosphere, the air in this layer cools as it flows sideways as well. Cool air descends at higher latitudes but eventually moves back toward the Tropics as the air cools and gets denser and descends more slowly. Warm air generally moves at high altitudes from the Tropics toward the poles. Low-altitude cool air flows backward from the poles toward the equator. This basic model of global air circulation was first proposed in 1686 by the English physicist Edmund Halley(Ćurić & Spiridonov, 2023). In the 1750s, George Hadley made modifications to the concept after realizing that the Earth rotation would change the direction of airflow. Taiga weather and climate are influenced by air masses. An air mass is an area of air that has been affected by the land or sea it is traveling over for a substantial amount of time to develop distinguishable features, including a specific range of temperatures and humidity.

Continental air masses are drier than oceanic air masses. Due to the fact that landforms cool more quickly than oceans, continental air masses are typically dry and cold during the winter. Since land masses warm more quickly than surrounding seawater, the continental air masses are likely to be both hot and dry throughout the summer for a given latitude. Heat is mostly transported from the equator toward the poles by ocean currents. The dominant winds, such as the mid-latitude westerlies, are the primary forces behind the currents. But the Coriolis effect, which causes the current to turn, affects the motion of water as it goes north or south. Because of this, the primary current systems in ocean basins often form gyres, which are circular systems. The current systems inside mid-latitude gyres flow clockwise in the Northern Hemisphere. The mid-latitude gyres are fed by mini-gyres that circulate counterclockwise, bringing colder water from the polar regions to higher latitudes. The air in coastal regions is cooled by currents traveling southward at the latitudes where taiga is developed. Compared to the continental interior, this impact typically extends taiga farther south in coastal regions. The northern edge of the taiga is moved farther north by the North Atlantic current, which also warms the air in northwest European coastal areas(Pielke & Vidale, 1995).

The North Atlantic current has significant consequences. Scientists studying climate change are worried that if global warming keeps up, it would stop the North Atlantic current from separating from the mid-latitude gyre. If the warm ocean circulation in northwest Europe disappeared, the region's average temperature would drop by at least 10°F (6°C). The current path of the North Atlantic current is maintained by cold saltwater sinking to the seafloor to form North Atlantic Deep Water (NADW) under the edge of the Arctic ice. To replace this water, surface water flows across from the North Atlantic current. If the creation of NADW is hindered by global warming, the North Atlantic current may not work. Researchers studying the remnants of zooplankton, or animal plankton, in the North Atlantic sediments have found layers of cold-water species sandwiched between layers of warm-water forms. The NADW appears to have been disturbed about 11,000 years ago, allowing cool-water zooplankton to flourish in the northeast Atlantic waters that are now home to warm-water plankton. Many plants and animals that live on the surface benefit from the snow. The snow does not quickly absorb the heat energy and melt as it would if it were darker since it is pale and reflects sunlight. Snow creates an insulating layer because it contains trapped air, which is a poor heat conductor. It lessens temperature changes and shields the soil from the cold air to keep it warm during the winter. At 20 inches (50 cm) below, the soil is normally at least 27°F (15°C) hotter than the surrounding air temperature in the middle of winter. When the snow layer is removed, the ground temperature drops sharply, increasing the likelihood that the surface soil will freeze. Many taiga plants are unable to withstand the rough winter months without a layer of snow, but some have developed special adaptations that allow them to stay above freezing even in the absence of an insulating layer of snow.

The majority of fungi are only visible when their fruiting bodies, which are mushrooms, emerge from the forest floor. The network of hyphae that fungi have under the soil is a web-like structure. Numerous soil fungi associate with trees to form mycorrhizae. Boreal forests would not be nearly as fruitful as they are currently without mycorrhizae. Because conifers are hard and resistant to deterioration, it may take many years for bacteria and other decomposers to act in the cold forest floor of the taiga. The process of degradation is accelerated by mycorrhizae. In the soil surrounding conifers, mycorrhizal fungi grow their hyphae. The plant materials are broken down by the digestive enzymes they release. Through their intricate web of hyphae, the fungi transfer some of the nutrients to the roots of trees while absorbing the rest for their own purposes. A portion of the hyphae enter root cells and exchange materials with the tree. The hyphae system functions as an extension of the root system, providing the tree with a supply of newly recycled nutrients. They create antibiotics, which either eradicate or stop soil microbes that could endanger the tree. The tree provides a consistent supply of dead plant material, which is beneficial to the mycorrhizal fungi. In addition, the sugars and vitamins the tree provides the fungi with aid in their growth.

Mosses can only survive in moist conditions, yet they can tolerate partial drying during the year(Green et al., 2011). More than two-thirds of the water content of many moss species can be lost. When it comes to sexual reproduction, mosses do require plenty of water. Sperms from male reproductive organs known as antheridia are released into the water and swim to the female reproductive organs known as archegonia, which are found on the same plant or in other mosses of the same species. Compared to mosses, conifers and flowering plants produce pollen grains that are resistant to drying. Pollen does not need ambient water for the male gamete to successfully transfer; it is delivered by the wind or by insects. Unlike moss sperm cells that swim freely, their male sex cells may travel great distances and they do not require wet conditions. A structure known as the sporangium, or capsule, emerges from a stalk at the apex of the moss plant after successful fertilization. The capsule bursts open in arid weather, releasing spores that land and develop into new moss plants. Unusual symbiotic relationships between fungi and single-celled algae are known as lichens.

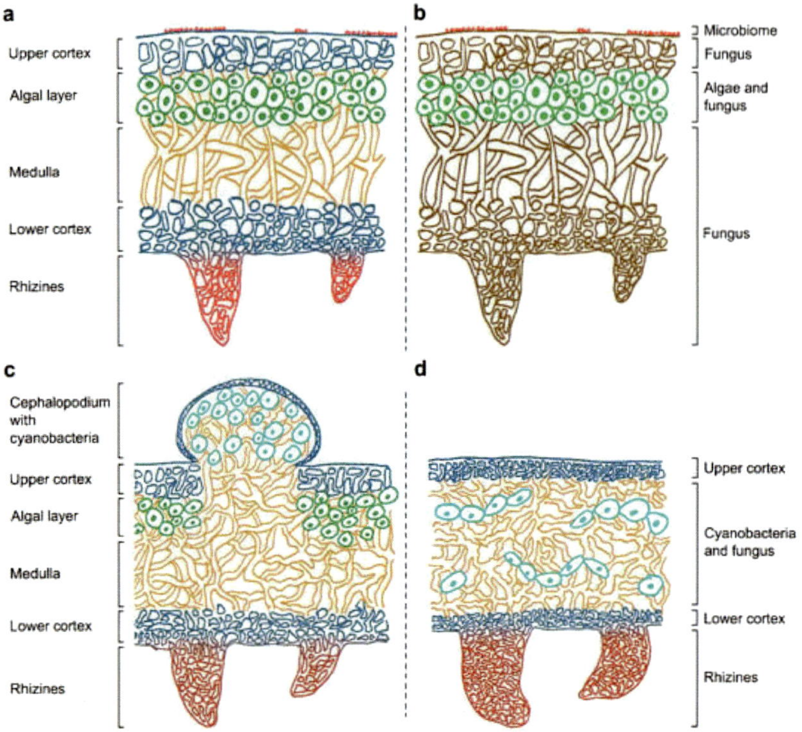

Figure 5.6 Lichens

Source: Morillas L, Roales J, Cruz C, Munzi S. Lichen as Multipartner Symbiotic Relationships. *Encyclopedia.* 2022; 2(3):1421-1431. https://doi.org/10.3390/encyclopedia2030096,*Creative Commons License.*

Lichens can be found growing on rock outcrops, tree bark, and the drier areas of the forest. Lichens have the same ability to dry and go dormant, only to grow back when they get moist. In addition to photosynthesis, the algal partner feeds the two (Figure 5.6). Both receive water and minerals from the fungal partner that are absorbed from snow and rain. Additionally, the fungus envelops its algae in a protective coating. The sheath contracts and shrinks in dry situations to stop water loss. The sheath expands and opens to permit the algae to resume photosynthetic activity when moisture is present again. Certain lichens that grow on trees yield acids that have antifungal and antibacterial qualities. All conifers and the majority of broad-leaved plants rely on wind for pollination. Most herbaceous plants found on the understory are pollinated by insects. They produce nectar-rich flowers, which draw insects.

5.10 FLORA AND FAUNA OF THE TAIGA

Small plants that depend on animals to disperse their fruits and seeds may have difficulties because the summer season is short. Most taiga shrubs and small flowering plants can reproduce both sexually and asexually as a safety measure against the failure of sexual reproduction. They produce overground or underground stems that extend laterally to form the roots and shoots of new plants. Many local animals primarily eat the fruits and seeds produced by taiga trees and bushes. Many plants have developed fruits that attract animals and birds to eat them, facilitating the plant species distribution. Ripe berries from juniper, bilberry, and other plants are eaten by a diverse range of taiga species, including brown bears, stoats, squirrels, and capercaillie birds. The seeds in the fruits frequently pass through the digestive tract undamaged, and the seed is sowed with a little bit of organic fertilizer when the animal excretes. Not only do animals and birds scatter seeds, but they also eat them. The tiny fruit of the twinflower shrub gets attached to the fur or feathers of animals. The taiga of Eurasia contains many of the related species of many animal groups found in North America(Fedorov et al., 2008). The taiga is home to both brown bears (*Ursus arctos*) and gray wolves (*Canis lupus*), which are sporadically spread from Alaska in the west to Siberia in the east. The most reasonable explanation for the presence of taiga species throughout a significant portion of the taiga zone which includes Eurasia and North America divided by the Atlantic and Pacific Oceans is the geological land connections that bind the taiga region together.

The most common explanation for regional variations in taiga animal species is the sporadic introduction of species from temperate or even tropical regions further south. Thus, musk deer that originated in Southeast Asia are found in eastern Siberia. The striped skunk and rufous hummingbird of North American taiga are closely related. The air near bogs and swamps can be populated with gnats, blackflies, and mosquitoes that feed on human blood. Both gnats and mosquitoes are members of the Culicidae family of insects, and many of them prefer the blood of mammals. The spruce budworm moth (Choristoneura) caterpillar in North America eats weak trees in a manner akin to that of the pine moth caterpillar. The temperature of their surroundings has a significant influence on the body temperature of reptiles and amphibians. They use behavioral strategies to regulate their body temperature, such as hiding in burrows during periods of intense heat to get their body temperature up to a point where they can become active. Few reptile and amphibian species can withstand the harsh winters in the taiga.

Figure 5.7 Siberian salamander

Source: https://commons.wikimedia.org/wiki/File:Salamandrella_keyserlingii_on_moss.jpg,*Creative Commons License.*

The Siberian salamander (*Salamandrella keyserlingii*), (Figure 5.7) can stay active at temperatures close to the freezing point of water, which is 32°F (0°C). The salamander hibernates during the winter and can endure temperatures as low as –31°F (–35°C) underground. The salamander lives on land for the majority of the year, and during the brief summer, it searches for a tiny pool of water to mate/breed. The male is the first to arrive and attracts the females by releasing chemicals known as pheromones into the water. The male fertilizes eggs after the females release their eggs into the base of the pond. After the eggs hatch, the larvae develop in the water before migrating to a safe place for the winter on land. In the summer, temperatures can fluctuate so much that eggs laid by reptiles on the ground have very little chance of surviving. The two common reptile species of European boreal woods, the European viper (*Vipera berus*) and the lacertid lizard (*Lacerta vivipara*), avoid this risk by giving birth to their young. The majority of the bird species found in the taiga were once passerines, or perching birds, which evolved in temperate woodlands. In winter, the majority of these birds fly south from the taiga, and they return in summer.

Chapter 6

INTRODUCTION TO GRASSLANDS

Contents

6.1INTRODUCTION

More than 40% of the Earth's surface was covered with grasslands. Across the globe, grasslands constitute one of the ubiquitous types of vegetation, and grasses are the most widely distributed group of plants. They can be found on every continent except Antarctica. A sizable portion of grassland has been turned into agricultural land. Wheat and corn grow well in areas with enough rainfall because of the abundant sunlight and rich soils. Pasture grasses that have been sown for animal feed can also be found in agricultural grasslands. In temperatures too dry for most trees, grasses grow well. They can withstand animals grazing on them, but they cannot withstand shade. As a result, natural grasslands are found in regions of the planet that are neither too shady nor too dry, like deserts or forests. The main types of grasslands are tropical savanna grassland and temperate grassland. The most common locations for temperate grasslands are in North America, Eastern Europe, Turkey, northern Iran, central Asia, and a portion of southern China. Moderate grasslands are found in areas with temperate climates. Southern temperate grasslands are limited to a small portion of Argentina, Peru, and South Africa since there is less land in the Southern Hemisphere, south of the Tropic of Capricorn(Eidt, 1969). Only patches of tropical forest separate the majority of African savanna grassland, which stretches from the southern edge of the Sahara to the southeast corner of the continent. Large savanna regions can also be found in Australia and South America. The same climate that supports grasslands can also be found in mountainous areas. There is a certain altitude at which the temperature drops to the point where trees cannot withstand the cold temperatures of the mountain climate. The growing season is so short at this height that coniferous tree needles cannot fully develop before being damaged by frost. While many flowering herbs and grasses can thrive, trees cannot. Mountain grasslands, often referred to as alpine meadows, are found between the area of bare rock and permanent snow and the tree line. They differ in composition from place to place, and alpine savannas can also be found in the Tropics.

6.2 GRASSLANDS

Alpine grasslands are found worldwide in mountain ranges. After other vegetation has been removed from the land, grasses quickly take over. Usually, as higher plants like trees and shrubs take root and shade the ground, these grasses will disappear. However, in certain areas, this has not occurred, leading to successional grasslands that are managed by grazing or burning, both of which damage the seedlings of shrubs and trees. Tropical forests have been replaced by successional savanna grasslands. The climates of tropical and temperate grasslands differ. Because climate mostly depends on distance from the equator and the ocean, they are limited to specific latitudes. Different plant species thrive in various climates. Although grass is the most noticeable and prevalent plant in all grasslands, there are many different species of grass, and some species cannot live in other types of grasslands. There are several traits that all grasses have in common. However, there is a significant difference between temperate and tropical grasslands due to the diversity of grasses that give each type of grassland its unique appearance. Temperate grasslands are plains covered in grasses and herbs. Most savanna grasslands (Figure 6.1) support forest plants, such as trees and shrubs, while other regions of tropical grasslands are covered in grasses and herbs. While the amount of forest plants varies from location to location, they are responsible for much of the savanna's transformation into an open forest, where grass surrounds trees and shrubs.

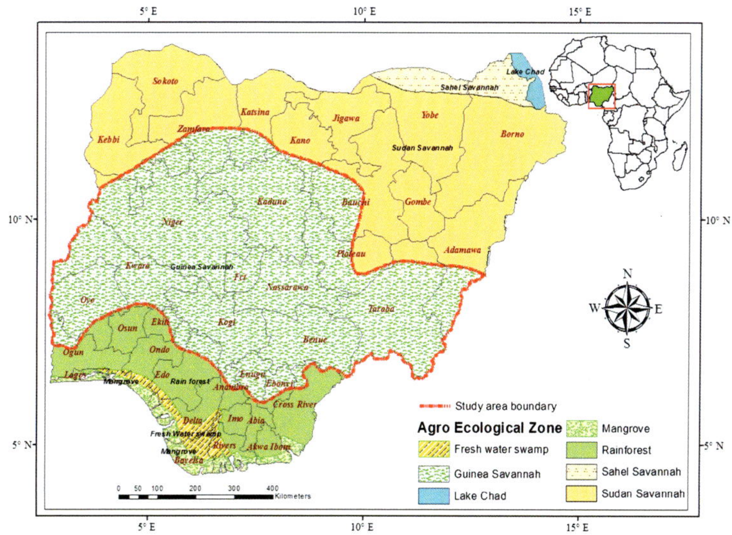

Figure 6.1 Savanna grasslands

Source: Adenle AA, Ifejika Speranza C. Social-Ecological Archetypes of Land Degradation in the Nigerian Guinea Savannah: Insights for Sustainable Land Management. *Remote Sensing*. 2021; 13(1):32. https://doi.org/10.3390/rs13010032, *Creative Commons License.*

Everywhere in the savanna, there is constant heat. While it occasionally dips as low as 40°F (4°C), the average annual temperature is consistently higher than 64°F (18°C). In the summer, it reaches 80°–100°F. Rainfall varies during the rainy season; it can be as low as two inches in the desert or as much as 60 inches in the tropical rainforest. Heavy rain frequently results in floods in some areas. Where there is little rainfall, savanna grassland predominates, whereas savanna forest is found in the wetter regions. Additionally, temperate grasslands thrive in a climate that is generally dry with occasional extended droughts. Occasionally, droughts can endure for ten years. On average, the grasslands of North America receive 40 inches of rainfall annually in the east and 10 inches (250 mm) in the west. The majority of this precipitation occurs during the winter months. The summers are not only hot and dry, with highs of 100°F. The constant wind and high summer temperatures cause a high rate of evaporation, which leaves the soil extremely dry.

Figure 6.2 *Andropogon gerardii*

Source: https://commons.wikimedia.org/wiki/File:Andropogon_gerardii_(3904160434).jpg,*Creative Commons License.*

In tallgrass prairies, *Andropogon gerardii*, (Figure 6.2) or big bluestem, is the most common type of grass(Scott & Baer, 2018). When it reaches an average height of five feet in late summer, it initiates flowering. The giant bluestem can grow up to ten feet tall in certain areas. Additional tallgrass species include switchgrass, Indian grass (*Sorghastrum nutans*), slough grass, and sand bluestem. A vast variety of herbs known as forbs grow among the grasses. In the east, tallgrass prairie is found. Summer rainfall predominates over winter rainfall, with an average of 34 inches per year. The native prairie used to be extremely moist in some areas in the spring, after the thin layer of snow had melted but before the temperature had increased sufficiently for water near the surface to evaporate. Now that the area has been plowed, wheat and corn are being grown there. The grassland known as short-grass prairie expands to the west of longitude 100°W due to the dry environment. For instance, the average annual rainfall in Great Falls, Montana, is 15.3 inches. Only in areas that have access to irrigation can farmers cultivate row crops; otherwise, the land is primarily used for cattle grazing. Additionally, plants are unable to grow very tall due to the less rainfall.

6.3 GRASSES

The majority of grassland plants are hardly taller than eight inches. These include several grama grasses (Bouteloua species), Pennsylvania sedge, galleta grass (*Hilaria jamesii*), buffalo grass, and needle-and-thread grass. Short-grass prairies have fewer forbs than tall-grass prairies because grasses are better in this climate than other plant life. The mixed prairie, which lies between the tallgrass and shortgrass grasslands, is made up of grasses. Although the majority of the region is now farmed, with corn and soybeans being grown, tall grasses flourish in the areas with water. The tallest grassland plants reach a height of almost three feet. They consist of June grass (*Koeleria pyramidata*), wheatgrass, wild rye, needle grasses, dropseed grasses (Sporobolus species), and little bluestem grass (*Schizachyrium scoparium*). Cultivated rye and wild rye are not related.

In Washington State, the Palouse River is the center of the Palouse plains, which stretches east and north. It is mostly covered in bunchgrasses. The majority of prairie grasses have roots that develop continuously below the surface and periodically extend new shoots from their horizontally growing stems. Branches split from the base of the main stem in bunch or tussock grasses. Palouse grassland grasses like bluebunch wheatgrass are common there. California bunchgrass grassland has disappeared. Mid-19th-century travelers described it as a 400-mile stretch covered in an excess of purple and yellow flowers in the spring. The primary grass was Stipa pulchra, or purple needlegrass. The grasslands of Europe and Asia are the largest in the world, spanning 3,500 miles (9,000 km) to the east from the Danube to Mongolia and China, and from the southern fringe of the Arabian Desert and the Sahara to the northern taiga.

6.4 STEPPE

Occasionally, temperate grasslands of all kinds, such as veld, pampa, prairie, and Australian grasslands, are referred to as steppes. However, there are distinctions between these kinds; in technical terms, steppe refers to the grasslands of Eurasia. With the exception of the point where the steppe transitions into the northern forests, these are wide, grass-covered plains devoid of trees. The majority of the area experiences a dry environment with 12–20 inches (300–500 mm) of rain per year, and droughts are frequent. Drought is not a threat to any steppe plant. Shorter plants grow where the earth is mostly dry, and taller plants grow in moist locations. Feather grasses are the common grasses in the dry southern steppe, while many other types of grasses also flourish in this large area. As their name implies, they bear long, feathery flowers that, by the end of May, cover the landscape. The genus Stipa, which includes feather grasses, may have derived its name from the Latin steppe. There are also abundant species of fescue (*Festuca* species), sedges (*Carex* species), and oat grasses (Avena species). Plants like *Poa bulbosa*, Gagea species, and tulips (*Tulipa* species) are found in areas with warm, moderately humid climates in Kazakhstan and the southern part of the region, close to the Black Sea. Wormwoods (Artemisia species), also known as sagebrushes, are found in the steppes of Central Asia. There is diverse vegetation in the steppe. In certain areas of the northern steppe, there are as many as 73 different plant species per square yard of surface area(Lamprey, 1964; Lisetskii et al., 2016).

Russian scientists have classified the steppe, which is located between latitudes 40°N and 50°N, into categories of grasslands. The taiga is bordered by the northern steppe. We refer to it as forest steppe. Trees are sporadically found all throughout the place, with the frequency increasing in the north until the steppe blends into the coniferous forest. The meadow steppe, located just south of the forest-steppe, experiences a wet environment. There, feather grasses and sedges grow. A portion of the meadow steppe is utilized for cattle grazing. South of the meadow-steppe is dry steppe. Because there is less annual rainfall than the amount of water that may evaporate in a year, the ground is typically dry there. Fescues and shorter feather grasses like *Stipa ucrainica*, *Stipa pennata*, and *Stipa capillata*, known locally as tyrsa, grow. As we move south, the steppe transitions into the subtropical desert, bringing with it an increasingly drier environment. All four varieties of steppe share a similar climate, despite the north having a wetter environment. The winters are rough, with at least four months of below-freezing temperatures and more than four inches of snow on the ground. The latitude of Irgiz, Kazakhstan, is 48.37°N, which corresponds to average steppe temperatures. From November through March, the average temperature stays below freezing. The average summer temperature is 73°F (23°C), and it typically lasts from June to August. Nonetheless, Irgiz experiences relatively little rainfall annually, just seven inches on average. Although there is an equal distribution of rainfall throughout the year, there are typically just 11 days of rain between May 1 and August 31. The longest days of sunlight occur throughout these months, and any rain that does fall soon evaporates. The steppe stays dry until the next spring by August when the grasses are all brown. Late summer and early fall are popular times for droughts. An open area is termed a bamba in Quechua, the Native American language spoken throughout most of western South America. The term was altered to pampa by Spanish colonists, who used la pampa to refer to a large plain. Although the singular pampa is more accurate, English speakers frequently use the plural version, pampas, to refer to the grasslands. Over 290,000 square miles (751,000 km2), primarily in Argentina but expanding into Uruguay, is the total area covered by the pampa. The majority of the pampa region is currently used for the ranching of sheep and cattle.

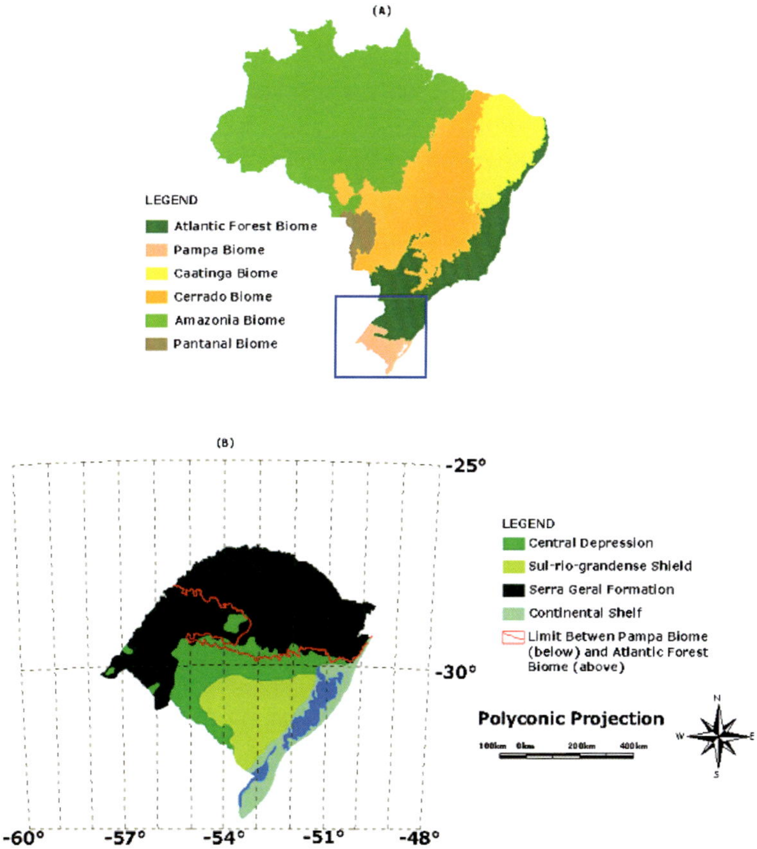

Figure 6.3 Pampa

Source: Roesch LFW, Vieira FCB, Pereira VA, Schünemann AL, Teixeira IF, Senna AJT, Stefenon VM. The Brazilian Pampa: A Fragile Biome. *Diversity*. 2009; 1(2):182-198. https://doi.org/10.3390/d1020182,*Creative Commons License.*

6.5 PAMPA

The Pampa (Figure 6.3) is a region that stretches from the Atlantic coast in the east to the Andes in the west, and from the Colorado River in the south to the Gran Chaco(Berón, 2016). The pampa is located roughly between latitudes 27°S and 38°S, and it experiences hot weather. In summer, temperatures typically range from 85°–90°F (29°–32°C), whereas in winter, they range from 57°–59°F (14°–15°C). On the other hand, winter nights are cold, frequently dropping below freezing. A plain extends just above sea level to an elevation of 2,320 feet (708 m) in the Andean mountains of Mendoza Province is separated by hills in the south and highlands, or sierras, in the northwest, including the San Luis and Córdoba sierras. The sterile pampa is a dry plain on the western side that occasionally gives way to a sandy desert. Bahía Blanca in the south receives an average of 20 inches (500 mm) of rainfall annually, whereas Victorica in San Luis Province in the west receives about 22 inches (559 mm). The soils of the eastern pampa are fertile and get an average of 39 inches (1,000 mm) of rain annually. The annual average amount of rain in Buenos Aires is 37 inches (940 mm). Argentina's most productive agricultural land is found in this area of the pampa. Pampas grass is planted in many parts of the world but has turned into a highly invasive weed in other areas, such as California, Hawaii, and New Zealand. *Cortaderia selloana*, also referred to as Uruguayan pampas grass, is a dense tussock grass that grows eight to twelve feet (2.4–3.7 m) tall and bears big flowers. Except for the herbs that grow in between its tussocks, silver pampas grass throughout a significant portion of the pampa. It is not very useful for pasture; instead, it is utilized to create paper. The majority of it has been replaced with more nutrient-dense pasture grasses and plants.

The eastern pampa is devoid of trees because the powerful winds produce frequent winter fires that they cannot withstand. After a fire, the grasses bounce back fast and gain from the nutrients in the wood ash that the rain rapidly carries to their roots. The ombu (*Phytolacca dioica*) is the only plant that can resist fire. Its girth, or the circumference of its trunk measured approximately four feet above ground, can reach 40–50 feet (12–15 meters), and it can grow to a height of 40–60 feet (12–18 meters). The plant frequently grows several trunks that are arranged side by side. Despite its height, its structure technically classifies it as a shrub rather than a tree. Because the ombu trunk contains tissues that retain water, it is resistant to fire. It can withstand droughts due to its capacity to store water. In regions like Southern California with suitable climates, ombus trees are planted as shade trees. The eastern part of South Africa is made up of a high plateau covered in grasslands known as the veld. From sea level to over 9,000 feet, the elevation changes. The major vegetation in most of the llanos is grass. For grasses, the most common genus is Trachypogon. Trachypogon grasses reach a height of over six feet when they grow in tussocks spaced four to twelve inches apart Herbs growing up to three feet tall can be seen in between the tussocks. There are a few different grasses growing on the llanos. These include a variety of palm trees; the golden spoon (*Byrsonima crassifolia*), which yields edible fruits; *Curatella americana*, whose leaves are used to polish metal; and *Bowdichia virgilioides*, whose bark is used to treat rheumatism and tuberculosis.

The most common type is called *Themeda triandra*. It can reach a height of 12–36 inches (30–90 cm), however, it is easily overgrazed and has low nutritional value. Love grass (*Eragrostis* species) frequently replaces red oat grass in areas where it has been overgrazed(Voigt et al., 2004). Bermuda grass grows to a height of four to fifteen inches (10–40 cm), while bristle grass reaches 20–36 inches (50–90 cm). The wet environment of the higher land is ideal for the growth of bluegrass. The veld has long been used for agriculture. The majority of South Africa's dairy, beef, and wool products are sourced from it, and large portions have been turned into crops. The most significant crop is corn or maize, however, sorghum, wheat, and sunflowers are also farmed. A portion of the veld is heavily industrialized and urbanized. Johannesburg and Pretoria are situated in the Witwatersrand, a substantial mining and industrial region. The majority of us imagine tropical South America, which is located east of the Andes, as a region covered with forests that thrive in a humid, hot climate. Open grassland makes up about half of tropical South America instead of any kind of vegetation. The term llanos refers to the grasslands in Venezuela, which span roughly 125,000 square miles (323,750 km^2), and the term eastern llanos refers to the grasslands in Colombia, which span 100,000 square miles (260,000 km2). Plains bordered to the north by the Cordillera Mérida Mountains and to the south by the Orinoco and Guaviare Rivers are covered by the llanos grasses. The llanos were first exposed to cattle in 1548, and today it is an important area for breeding and ranching. Situated between the Guiana Highlands in the east and the Andes in the west, the vast basin where the llanos are located was created millions of years ago and then filled with silt. Subsequently, the basin sank in certain areas, resulting in the nearly level plains that are broken up by mesas. It is likely that the grasslands are only 10,000 years old.

The weather in Cerrado is tropical. With an average maximum of 81°F (27°C) in the N and 72°F (22°C) in the S of the region, it is hot year-round. The summer dry season spans five or six months over two-thirds of the Cerrado, with some months experiencing no rain at all. The tropical grasslands of Central and South America, as well as the Caribbean islands, were once referred to as savannas or savannahs. Nowadays, the term refers to any tropical grassland with temperatures that never drop below 64°F (18°C), a rainy season, and a winter dry season that lasts three to five months. However, the term savanna is now particularly linked to Africa. The African savanna's northern boundary stretches from the southern limit of the Sahara to the Nile River, following a path that passes through Senegal and Guinea-Bissau. The grassland extends southward to the Ubangi River and Lake Victoria in Central and East Africa, as well as the border of the tropical rainforest in West Africa. The Namib and Kalahari deserts are not included in the savanna, which extends south of the equator to the southern boundary of the tropical forests in South Africa, or latitude 7°S to latitude 32°S. It covers an area of almost 5 million square miles. That is larger than the entire size of the United States, and it makes up about half of all of Africa. The majority of the region is categorized as savanna. This indicates that it was formerly a forest, but that most of the trees were removed a long time ago. The cleared land was filled with grasses, and then herds of grazing animals flattened the tree seedlings, preventing the forest from returning. In addition to occurring naturally throughout the dry season, fires keep the forest from recovering. There are grasses in its place. The tall grasses are found in the wetter regions, such as the edges of tropical forests, with an annual rainfall of 24 to 40 inches(Janzen & Schoener, 1968). Hyparrhenia species reach heights of 10–13 feet. One species, *H. filipendula*, is farmed in some savanna areas specifically for paper making.

6.6 THORN TREES

The most recognizable woody plants are thorn trees (Acacia species), of which *A. tortilis* is the most common. It can reach a height of 13 to 50 feet (4 to 15 meters) and has the characteristic umbrella shape. The main source of gum arabic, which is used to create pastes and glues and is a part of various medications, is *A. senegal*. Another common savanna thorn tree, *A. laeta*, is a significant commercial supplier of gum arabic. The baobab (*Adansonia digitata*) and candelabra trees (*Euphorbia candelabrum*) are two of the most remarkable savanna trees that exist. The baobab can reach a height of 40 feet (12 meters), although its lower trunk can reach a diameter of 30 feet. Baobabs have an extremely long lifespan. The tree doesn't always grow; occasionally, it gets smaller, most likely due to fluid loss during dry spells. Every section of the tree has a purpose: Its inner bark is used to make rope, and its outside bark is used to make cloth. Its leaves are used as a vegetable, and its seeds are edible and high in vitamin C. Further, spurges and candelabra trees are closely related. The former is found in the savanna of East Africa. From the top of the stem, it has several branches that all point upward. Australia used to be a country full of lakes, rivers, and forests. A dry spell occurred approximately 10 million years ago. The trees dwindled along with the lakes, which disappeared. Grass may survive on less water than trees, but trees require plenty of it. Hence, the growth of grassland coincided with the withdrawal of the trees. Nowadays, a large portion of Australia's interior is desert. While there is some eucalyptus forest in the east and some tropical forest in the north, the majority of the continent is covered in a mixture of scrub and grassland for natural vegetation. Dwarf eucalyptus dense thickets can be found in the southern region of the country. Although it is made up of diverse species, the Australian grassland creates a savanna environment that is comparable to the African savanna(Young, 2020).

Figure 6.4 Red bloodwood

Source: https://commons.wikimedia.org/wiki/File:Corymbia_gummifera_habit.jpg,*Creative Commons License.*

Mitchell grasses can reach heights of 12 to 36 inches (30 to 90 cm) and occasionally 48 inches (1.2 m). They offer nourishing pastures and are drought-resistant. Cattle and sheep graze on them. The yearly rainfall ranges from 14 to 30 inches (350 to 750 mm). In summer, temperatures can reach over 100°F (38°C), yet in winter, frosts can occasionally occur. In the north, bluestem grasses replace Mitchell grasses. In dry climates, these grow to a maximum height of six inches; in more humid climates, however, they can reach up to 6.5 feet in height.

The Mitchell grass fields divide regions of forests where thorn trees predominate. More species of Acacia exist in Australia than on any other continent. The two most prevalent types found adjacent to Mitchell grassland are gidgee (A. *cambagei*) and brigalow (*A. harpophylla*). A vast savanna containing brigalow covers the majority of the Australian eastern region. It extends roughly from Townsville, Queensland, at latitude 19.25°S to the Queensland–New South Wales boundary, at latitude 28.58°S, encompassing 132,200 square miles. In this area, the annual rainfall varies from over 30 inches in the east to under 20 inches in the west. In dry regions, gidgee takes the place of brigalow, and in certain locations, there are open grasslands with a prevalence of bluestem grasses. Most of them are species of Eucalyptus.

Red bloodwood (*Corymbia gummifera*) (Figure 6.4) and spotted gum (*Corymbia maculata*), both of which are varieties of eucalyptus, can be found further south. There are areas with exposed brigalow forests. Mulga scrub, which is made up of Spinifex species of grasses and *Acacia aneura*, is found throughout the Australian western region. There is also mulga with saltbushes (Atriplex species, particularly A. vesicaria) in the center. On the southern side of the desert, savanna grassland with saltbushes covers a large portion of the continent. Saltbushes can grow anywhere there is saline soil. Mulga reaches a height of 20 feet. Its leaves are fed to cattle, and its wood has numerous applications. Mulga wood, for example, is typically used to make boomerangs. Rather than growing in tussocks, spinifex grasses are 20–40 inches tall and originate from underground stems. They are therefore highly helpful in stabilizing sand dunes (Figure 6.5) that are loose.

The plant composition of the forest progressively shifts between the densely populated area and the forest boundary. When broad-leaved evergreen trees initially appear, they may be followed by deciduous trees and finally coniferous trees like firs, pines, and spruces. At that point, the coniferous forest opens up considerably. The trees are widely spread, with gaps where sunlight reaches the forest floor, allowing grasses and herbs to thrive, as opposed to being densely packed, which would have made the forest floor extremely dark. The trees get smaller and distorted as they disperse more and farther apart. This is the tree line, when there are finally no trees at all. It is more like a band where the concentration of trees progressively diminishes than a well-defined line.

Figure 6.5 Sand dunes

Source: https://commons.wikimedia.org/wiki/File:Dyyni_Yyteriss%C3%A4.jpg,*Creative Commons License.*

6.7 TEMPERATURE ALTERATIONS AND TREES

As one moves farther from the equator, typical summer temperatures drop, and temperate grasslands replace forests when these temperatures fall below 50°F. However, mountains are more frequently referred to as the tree line. There is a height at which trees cannot survive because the temperature drops with altitude. Some meadows form alpine savanna in the Tropics and alpine grassland elsewhere above this tree line, whereas the forests that cover a lower slope gradually become sparser and eventually disappear. The environmental lapse rate (ELR), which varies, is the rate at which temperature drops with height(Jobst et al., 2017). However, since the average summer temperature is what decides whether or not trees survive, it is the average ELR that counts. We may determine the approximate height of the tree line if you know the location height above sea level as well as its normal summer temperature. In El Paso, Texas, for instance, the warmest month average temperature is 80°F. At an elevation of roughly 8,300 feet, the temperature will drop by 30°F to reach the transition point of 50°F. El Paso is already 3,920 feet above sea level, thus at an elevation of 12,220 feet, the average summer temperature will be 50°F. As a result, on the mountains above the city, that will be the tree line average height.

There are sections of the mountain that face the sun, some that face away from it, and some sections that will receive full light and some that will receive partial shade. As a result, the average temperature on the slope will differ significantly from location to location. The temperature will frequently drop down the slope due to cold air from higher up, possibly from the area that is always covered in snow. This will lower the tree line. Alpine grasslands or savannas exist above the tree line, regardless of their elevation. Although the grassland composition varies, they are consistently rich in herbs and frequently include heathers growing in among the grasses. In most mountains, fescue grasses are prevalent. They are referred to as bunchgrasses or tussock grasses because they grow in clusters. Grasslands still exist in the interior of continents, but continents are shifting. Grass and related herbs do not grow on bare rock; rather, they grow in soil. Plant and animal remnants mixed with mineral fragments from rocks make up soil. When bare rock is exposed to the elements, wind and water erosion breaks it down into tiny fragments. We refer to this process as physical weathering. Plants have specific climatic requirements, and climates vary widely. The llanos and cerrado grasslands of South America are primarily made up of soils that have lost the majority of their plant nutrients(Ayarza et al., 2022). The laterite layers that give them their reddish or yellow color can be found in many areas. African savanna soils are far more rich and younger. Poor, depleted, lateritic soils typical of tropical rainforests can be found on the southern side.

When the temperature is right for farming, the deep, productive soils of temperate grasslands, the prairie, steppe, pampa, and veld make excellent agricultural soils. Tropical grasslands occur in climates with wet and dry seasons. Grasslands flourish in climates that are too dry to support trees. Grasslands often receive a lot of rain. Originally prairie grasslands, the Great Plains of the United States are known for their storms; similarly, other grasslands throughout the world are subject to severe storms. The ground dries quite fast once the storm passes and the sky clears, despite the intensity of the rain. The water completely evaporates. Soils in grasslands are often well-drained. Vegetation covering the soil surface improves drainage of the soil. Because of the impact of heavy rain on exposed soil, water frequently remains on the surface of bare soil for a longer time after a severe downpour. Large raindrops hit the earth with significant force and descend at a speed of roughly 20 mph in calm air. The impact of the rain smashes the soil crumbs at the surface, separating the individual soil particles. Normally, dry soil particles cling to one another to create crumbs.

These are dispersed to cover the surface in a layer. Rain continues to pound the soil, compacting the particles until the layer forms a waterproof skin or cap that keeps water from soaking through. After that, water condenses into pools on the surface and gathers in hollows. The water evaporates while it is there and returns to the atmosphere unusable to the plants. On the other hand, plants that cover the ground prevent raindrops from falling. As the plants are impacted by the rain, they bend and rebound, releasing the water and allowing it to fall gently onto the soil. Raindrops that are caught by plants are not strong enough to break up soil particles, so the water can seep through and drain away.

Since trees and shrubs are having seeds, killing young plants lowers the amount of potential seed sources in the future. The amount of seeds in the soil gradually decreases as they sprout, grow, and eventually die. There are therefore no young plants to replace the mature plants that produced the seeds. Grazing really helps grasses, which is why they flourish when the trees and shrubs die. Grazing animals gain from the grass they eat. Most of their young survive because there is more food available, and since there are more animals to graze, the woody plants are further repressed. The animals keep the grassland from turning into a forest once it is big enough to accommodate sizable herds of grazers. Additionally, fire keeps grassland intact. Tropical grasses wither away during the dry season, leaving behind dry grass that can catch fire from even the smallest spark. Fires are useful and common. When the first rain falls, the ash layer that is left behind, rich in nutrients is carried into the earth, eliminating the dead plant matter. The downpour produces a package of rich, nutrient-rich grass because there isn't a layer of dead grass to prevent the new growth. The likelihood of trees and plants dying in a fire is higher. By the time they grow shoots above ground, the grasses are thriving and the grazers are grazing, even if their seeds below ground are unaffected. Grassland maintenance may involve human involvement as well. They hunt animals with fire as a tool and are dependent on them. Large animals run from fire, and hunters can make use of this habit to make hunting more convenient. In order to prevent the animals from seeing or smelling the hunters, a number of them hide. Grass soon reappears when the fire has subsided.

A grass species appears simple. It has long, narrow blade-like leaves, a stalk, and roots. Its petal-less flowers yield a lot of pollen, which is carried by the wind, and different amounts of seeds. For instance, grass flowers are in different sizes. While some are large, others are small. Pampas grasses, for example, have large flowers, while uva grass (*Gynerium sagittatum*), which grows in South American tropical grasslands, has flowers that can reach up to 6.5 feet in length. Animals can benefit from Grass. Cattle, sheep, and rabbits are examples of grazing mammals that consume grass leaves. Both people and a large number of rodents and birds eat grass seeds. Grass includes wheat, rice, corn, barley, oats, millet, and sorghum. Grasses are relatively new, highly evolved plants. The earliest plants on Earth were most likely green, single-celled organisms known as algae that floated close to the surface of the ocean when life first began on Earth.

6.8 MODIFICATIONS IN PLANTS

About 450 million years ago, plants initially made their way onto land. Algae, whose cells are connected to form long filaments, are most likely the ancestors of terrestrial plants. This type of algae, known as blanket weed (Cladophora species), is still present in slow-flowing rivers. These types of algae were seen growing around lakes and marshes. It would take about 390 million years for the first grass plants to emerge. Plants underwent a great deal of change over that long time. Plants generated sperm and eggs in structures called gametangia, where the eggs were fertilized and developed into new plants known as embryos. As they migrated farther inland, they also developed a hard, waxy outer layer that helped them retain water. They are the liverworts, hornworts, and mosses.

Water and nutrients could enter the cells of the plants and move to the areas where they were needed as long as they were tiny. Eventually, though, certain mosses with tissues that could transport water and chemicals dissolved in it started to grow. Plants were able to grow larger as conducting tissues kept evolving until they were able to carry nutrients and water throughout the entire, larger plant. A plant known as Cooksonia was upright, had branches, and generated spores in structures at the terminals of its branches approximately 410 million years ago. It was a plant with vessels, a vascular plant. During this era, plants reproduced by spores, which are minute particles, frequently including a single cell, containing genetic material that, under the right conditions, will germinate into a new plant. This is also the method used by ferns and horsetails, along with mosses, liverworts, and hornworts. In vast swamp forests that existed 350 million years ago during the Carboniferous epoch, huge ferns and horsetails flourished. Rather than spores, they generated seeds. A seed is a microscopic plant with primitive roots and leaves, as well as a food source to help it grow, all safely encased in a hard outer layer. Because they were more adapted to dry conditions than spore producers, the seed plants thrived when the Carboniferous epoch ended, when the climate shifted and the wetlands dried up. The gnetophytes, a class of plants, cycads, or ginkgo, and conifers plants with cones like firs, pines, hemlocks, and spruces are among the first seed plants that evolved into gymnosperms. Gymnosperms were the first plants to generate seeds to shield a small but fully formed plant structure. These adaptations were crucial to their survival. Gymnosperms don't have flowers, and the only covering for their seeds is a seed coat. During the Cretaceous period, roughly 130 million years ago, the first flowering plants developed(Bond & Scott, 2010). The majority of flowering plants have means of dispersing their seeds, and they use their flowers to draw in animal pollinators, most commonly insects. Within the ovary that develops into a fruit are the seeds. After eating the fruit, animals either scatter the seeds in their droppings or discard them.

Chapter **7**

BIOMES OF FRESHWATER AQUATIC SYSTEMS

Contents

7.1 INTRODUCTION

Every body of water is related to one another because they all take part in the hydrologic cycle. This relationship may or may not be important for the biota, depending on the specific hydrologic mechanism and the geographical location. For instance, even though there might not be a direct relationship when river levels drop and the river recedes from the floodplain-wetland system, floodplains are regarded as wetlands and as a component of the river system. River organisms inhabit the floodplain and its wetlands during the flood period when the river and the floodplain merge into one. This could be crucial to the reproductive cycle of a specific organism. Simultaneously, the floodplain-wetlands ecosystem receives more nutrients, organic matter, and sediment from the flooded river. Although the two systems may be regarded as distinct biomes, they are closely related. Surface flows, typically in the form of river systems, feed lakes of all sizes. Additionally, the hydrologic network offers a channel for the flow of sediment, the exchange of nutrients, and the distribution of species. Lake tributaries are where fish can lay their eggs. For instance, the Williamson and Sprague rivers serve as the breeding grounds for several species found in Upper Klamath Lake in the northwest of the United States. Typically, many fish species spend their early life stages in the wetlands surrounding Upper Klamath Lake(Markle et al., 2009). The diverse range of freshwater aquatic situations presents a multitude of options for feeding, dispersal, reproduction, and habitation, which have led to the evolution and coevolution over many years. Phytoplankton have the ability to alter their density. Photosynthesis occurs at the surface during the day, and as a result, the phytoplankton gain mass. Any gas vesicles are crushed by rising cell wall pressure, which also leads to the formation of relatively dense hydrocarbon molecules, which raises density even further. The phytoplankton sinks as the total density rises. Gas vesicles can enlarge, intracellular pressure is released, and some mass is released from the cell during respiration. The organism thus loses density and starts to ascend once more. In tandem with the daily cycle of light and dark, many phytoplankton go through this cycle on a daily basis.

7.2 SOME BASIC DATA

Algae have also been found to regulate density by secreting a mucus layer that retains and absorbs water, increasing cell size while lowering average density. Other adaptations support the placements of phytoplankton and zooplankton in the water column. These organisms must remain near the surface because that is where their food is found. Some algae have spines; these are believed to be used to increase surface area and prevent sinking. A lot of zooplanktons and some algae can move in the water column by using their abilities to propel themselves. To move forward, these algae develop an appendage called a flagellum, which resembles a whip and may be turned or whipped around like a propeller in certain situations. Since they can propel themselves, the other water column inhabitants, fishes primarily, have control over their depth. The largest class of fishes, known as Actinopterygii have a swim bladder or gas bladder that they may contract or expand to alter their buoyancy and average density. They accomplish this by first filling the bladder with gas produced internally or with air that has been absorbed at the surface, and then they empty it. Other fish store lipids, which are less dense than water, to increase their buoyancy. Water viscosity is somewhat temperature-dependent; it reaches its maximum viscosity slightly above freezing, and the warmer the temperature, the lower the viscosity. The Reynolds number is one technique to measure the interaction that a moving object (fish) has with water. The force of inertia, which causes an object in motion to tend to remain in motion, and viscous forces, which resist movement in a fluid, are precisely balanced by the Reynolds number. An object moving through a liquid is slowed by viscous forces but is kept moving by inertial forces. When the Reynolds number is more than 1, inertial forces take precedence; when it is less than 1, viscous forces do.

The size of the object in question has an impact on the Reynolds number. The Reynolds number decreases with the size of the object trying to pass through the water. Microorganisms with the ability to move through water have developed a variety of inventive ways to do so. In fact, photosynthesis is a process of absorbing and storing solar energy. However, because water is not totally transparent, light is absorbed and scattered by water molecules, which causes light intensity to decrease with depth. The drop in light intensity with depth is exponential. Plankton is one of the dissolved and suspended materials in freshwater that increases absorption, scatters light, and shortens its penetration distance. Sunlight disappears almost entirely below 330 feet (100m), even in clear water. The euphotic zone is the area where photosynthesis can take place due to sufficient light(Letelier et al., 2004). It is the area of the water column from the surface to the depth at which just 1% of the light that enters the water still reaches the surface. Photosynthesis and respiration are almost equal at this light level. Any plant, including phytoplankton, can only survive in the euphotic zone. An instrument known as a Secchi disc is used to measure the water's capacity to conduct light in natural surface water bodies. The disc is lowered into the water, revealing a striking black-and-white pattern. The marking on the line it is attached to allows one to note the Secchi depth or the depth at which the disc disappears. The euphotic zone extends downward two to three times the Secchi depth.

When a solid dissolves, it separates into positively and negatively charged particles known as cations and anions, respectively. These particles are referred to as salts. For instance, sodium chloride, or salt splits into positively charged sodium ions and negatively charged chloride ions. Some essential plant nutrients, such as calcium, sulfur as sulfate (SO_4^{2-}), nitrogen as nitrate (NO^{3-}), and phosphorus as phosphate (PO_4), are also commonly found as ions from dissolved salts. An electrical current is often passed through water to measure the conductivity of the water over a fixed distance in order to ascertain the amount of dissolved salts present. Resistance has the opposite effect on conductivity. The conductivity of water is a useful indicator of the concentration of ions present since ions are good conductors while clean water is not. Temperature has an impact on water conductivity; therefore, measurements need to account for that. Microsiemens per centimeter (mS/cm) is the unit of measurement for conductivity; a low level could be as low as 50 mS/cm, while a high level, like that of saltwater, is 32,000 mS/cm. At 158,000 mS/cm, the Great Salt Lake in Utah records an extremely high value. The size and geology of the watershed, land use and human activities (agriculture) within the watershed, atmospheric deposition, biological processes in the lake, especially in the hypolimnion, and evaporation all affect the amount of dissolved salts in a lake. Because evaporation rates in dry places are higher than rates of precipitation, dissolved minerals in the water are concentrated in the evaporated water, which is why lakes in arid regions are frequently salt lakes, like the Great Salt Lake. Total dissolved solids are closely associated with dissolved salts. This represents the overall amount of dissolved solids in water of all types. The primary component of natural waterways is dissolved salts, however other substances, such as dissolved organic compounds and hazardous organic contaminants, may also be present. Conductivity actually just measures the concentration of ions, even though it is frequently employed to measure total dissolved solids. By evaluating the solid residue that remains after evaporating a certain quantity of water, a more precise measurement can be made. The majority of aquatic species face significant challenges due to elevated dissolved salt levels.

7.3 HOW CELLS REACT TO OSMOTIC PRESSURE

The majority of living cells use internal pressure on their cell walls to preserve their structural integrity. This pressure is produced in aquatic single and multi-celled organisms by an osmotic difference between the water surrounding the cell and its contents, which have a lower concentration of ions. Such an organism will dry if it is submerged in salty water because water molecules will osmotically migrate across the cell wall in the direction of the higher ion concentration. Moreover, inorganic ions enter cells through the cell wall and become hazardous when their concentration rises. One-celled organisms that use salt have evolved by keeping their internal ion concentrations higher, though maybe not exactly the same as what is found in the saltwater outside of their cells. It appears that the best tool to repel sodium ions is potassium ions. We refer to these species as halophilic. Certain organisms (facultative halophiles) are tolerant of high salt concentrations, whereas others really need high salt concentrations (obligate halophiles)(Oren, 2008). Similar cellular modifications are used by halophilic plants, or halophytes, to enable them to survive in salinized environments. Additionally, many contain specialized cells or organs that eliminate salts to stop them from building up, in addition to others that stop salts from entering the plant deeply, especially the roots in the case of developing vegetation. Controlling their ion concentration is a common adaptation of animals living in saltwater environments. Larger animals have specialized regulatory organs that control the salt content and eliminate excess salt for excretion, much like human kidneys do. However, very few species can withstand the extraordinarily high salinities seen in salt lakes. The brine shrimp, (Figure 7.1) which are crustaceans belonging to the genus Artemia, are one exception. These tiny organisms can survive in a range of salinities. Due to a lack of aquatic predators, brine shrimp can grow to huge populations in hypersaline situations. It is interesting to note that fish from both freshwater and saltwater maintain roughly equal internal salinity levels. Due to osmotic pressure, salts that are continually invading their bodies must be collected, transported, and expelled using specialized organs because salt water is around three times saltier than their blood.

Figure 7.1 Brine shrimp

Source: https://en.wikipedia.org/wiki/File:Artemia_salina_4.jpg,*Creative Commons License.*

7.4 THE IMPORTANCE OF PH

For many biological processes, pH is crucial. pH affects how soluble organic compounds, metals, and nutrients are, as well as how soluble organic compounds are. It is obvious that pH plays a crucial role in deciding the kinds of species that can exist in a body of water and the population levels of those organisms, since the availability of nutrients, like phosphorus, to organisms depends on their chemical form, which is regulated by pH. For instance, acidic waters decrease the bioavailability of calcium, which is required for the synthesis of bones and shells. On the other hand, raising acidity increases the bioavailability of metals and may be harmful. Fish are typically the most important lake species, and most fish can live at a pH range of 6 to 9(Rahel, 1984). Different fish species have varying tolerance ranges for alkalinity and acidity, and within a species, there are variations in tolerance during different life stages. There might be a difference between the pH that is acceptable for reproduction and survival. Aquatic animals and biotic communities have been affected by the acidification of freshwaters brought on by acid precipitation, also known as acid rain, and the consequences of mining. Lake pH varies and is influenced by several factors.

Water that dissolves in rocks like limestone and dolomite loses hydrogen ions, making the water alkaline. Lakes that receive water from streams and groundwater that permeate this type of sandstone are typically alkaline. In the face of both naturally occurring and contaminated acidic precipitation, these lakes will be able to withstand acidification. Because carbonic acid is formed when atmospheric carbon dioxide reacts with water, natural rainfall has a pH of about 5.6. Fossil fuel-related air pollution has the potential to significantly alter pH of precipitation, with some locations recording as low as 3. In certain areas, acidification of lakes is common. Certain lakes become acidic due to high levels of dissolved organic carbon from vegetation, whereas other lakes may become alkaline due to industrial pollutant discharge. pH can also be changed by biological processes. While respiration releases carbon dioxide (CO2) into water, lowering pH and increasing acidity, photosynthesis absorbs CO2 from water to increase pH. The pH of eutrophic lakes can change every day.

7.5 GAS DYNAMICS

Because gases also dissolve in water, surface waters that are exposed to the atmosphere will have dissolved gases that, while not exactly reflecting the chemical composition of the atmosphere, will nevertheless contain some of it. About 78% of the atmosphere is made up of molecular nitrogen gas (N2) and 21% is made up of molecular oxygen gas (O2). Smaller amounts of argon, carbon dioxide, and many trace gases are also present in the atmosphere. In aquatic systems, oxygen has a major biological role, but dissolved molecular nitrogen is not as important.

To release the energy contained in the carbon-based molecules that comprise their sustenance, the majority of living things need oxygen. This process is called respiration, and it is opposite to photosynthesis chemically. Plants use sunlight during photosynthesis to mix water and carbon dioxide to produce carbohydrates. During respiration, carbon dioxide and water are produced from carbohydrates by plants, animals, and other organisms using oxygen. This process releases energy. The temperature has an inverse relationship with the ability of gases, including oxygen, to dissolve in water: the higher the temperature, the lower the gas concentration that can occur; conversely, the lower the temperature, the higher the concentration. It is said to be at its saturation level when the dissolved oxygen concentration, expressed in milligrams of oxygen per liter (mg/L) of water, is at its highest point for that particular temperature. Liquid water just above freezing at sea level has the maximum saturation level of oxygen; this amount falls linearly with temperature. The saturation level is also influenced by elevation and pressure. Since atmospheric pressure drops with elevation, so does the saturation level at any given temperature. Biological activity is another factor that affects the oxygen content of water. All plants use carbon dioxide that is dissolved in water during photosynthesis to make oxygen. In water, photosynthesis results in an increase in oxygen content. Oxygen is used by plants, animals, and microorganisms for respiration, which lowers oxygen levels. Diffusion from the atmosphere replenishes oxygen that has been depleted by biological activity in the near-surface zone of lakes. However, oxygen diffuses slowly in water, thus to get oxygen down into deeper water, the water column must physically mix. While lakes lack turbulence, rivers always have an adequate supply of oxygen. Oxygen levels can be drastically reduced in the absence of active mixing. This is especially true in the dark, deep waters where photosynthesis is not possible but significant breakdown is. In areas where photosynthesis is present, oxygen levels may fluctuate on a daily basis. In response to these fluctuations, organisms have evolved diverse tactics to cope with them. For their survival, almost all freshwater aquatic species require oxygen. However, even in ideal conditions, the oxygen content of water is rather low when compared to the atmosphere. The dissolved oxygen values for the majority of fish species range from 7 to 9 mg/L(Miller et al., 2002). The oxygen content in cold, turbulent streams can range from 9 to 12 mg/L, which is close to the maximum value observed in freshwater habitats. Low oxygen levels can occur under less ideal conditions; fish and other species will then either migrate to areas with higher oxygen levels or die.

7.6 AQUATIC BIODIVERSITY

Aquatic life requires oxygen to enter their bodies. As a result, they lack specialized breathing organs and take in oxygen through their skin or cell walls. For small species like phytoplankton and zooplankton, this is an effective tactic since the body surface-to-volume ratio increases with size. Furthermore, when an organism is very small, oxygen diffuses naturally through its cells without the requirement for a specific mechanism. Water can dissolve oxygen, but oxygen does not diffuse through it as well as it does through air. As a result, an aquatic organism has the potential to exhaust the surrounding oxygen through respiration.

The species will be surrounded by a layer of low-oxygen water. Aquatic species have evolved a variety of strategies to maintain water flow and stop the formation of a water layer that is becoming depleted. This issue is resolved by multicellular species that use integumental respiration by moving or merely bending their bodies. Fish and other larger aquatic species have far more complex mechanisms in place to deal with their oxygen requirement. Some come into periodic contact with the surface to draw on atmospheric air. Some insects can breathe air without coming out of the water due to appendages.

Gills enable them to greatly increase the surface area over which oxygen will diffuse; by maintaining water movement across the gills through flapping a portion of their body, forward motion, or both. Gill structures are also utilized by a large number of aquatic insects and other macroinvertebrate larvae. Larger, multicellular plants known as macrophytes are essential to many aquatic ecosystems. Although they are rooted in the submerged substrate, developing plants rise above the water. Similar to this, floating-leaved plants have roots that float on the top without growing below it. As in many tropical freshwater habitats, free-floating plants are not rooted; instead, they float freely on the surface and frequently form huge floating mats upon which other kinds of life establish themselves(Den Hartog & Van Der Velde, 1988; Sasser et al., 1996).

Major aquatic macrophytes including angiosperms, bryophytes, and certain large types of algae are visible without a microscope. Numerous factors influence the distribution and quantity of macrophytes in freshwater aquatic settings. Because of depth and frequent turbidity, light availability inhibits the growth of some macrophytes in shaded streams and confines populations to the shallow edges of many larger rivers and lakes.

The establishment of macrophytes is frequently not facilitated by substrate conditions, and in aquatic habitats with low nutrient supply, nutrient availability may be restricted. A broad collection of unicellular species, including ciliates are known as protozoans. They typically favor locations with slackwater, sites where suspended material is deposited, and the spaces between particles of sand or gravel. While some protozoans are parasites, others are predators, and yet others feed on bacteria and algae. Rotifers are microscopic organisms, which are hardly noticeable to the unaided eye and are a significant component of the zooplankton in most freshwater settings, together with protozoans. For certain fish, they are a key source of food. A large genus of organisms known as rotifers eat bacteria, algae, and small animals. Flatworms/ Planaria (Order Tricladida) are the important members found in streams and rivers.

They are mostly found in headwater streams since many of them like cold water. Planaria are mostly 0.2–1.2 in (5–30 mm) long, flat or ribbon-like worms that glide over the substrate while scavenging or foraging for food. Nematodes are microscopic roundworms that lack segmentation. They are a very diverse and common group that live in freshwater, terrestrial, and marine situations. Some free-living organisms live in freshwater habitats and are parasitic. The two major groups of Annelida phylum that are found in freshwaters are oligochaetes and hirudinea. The majority of oligochaetes are detritivores or scavengers found in sediment deposits. While some leeches are parasites, others are scavengers. Since annelids can withstand low dissolved oxygen levels, their dominance suggests that the water quality is not good.

Figure 7.2 Sponges

Source: https://en.wikipedia.org/wiki/File:Aplysina_archeri_(Stove-pipe_Sponge-pink_variation).jpg,*Creative Commons License.*

7.7 SPONGES

Sponges (Figure 7.2) belong to Phylum Porifera and are often found in marine habitats, there are over 150 types of sponges found in freshwater settings(Meixner et al., 2007). They are simple morphologically, with sizes ranging from less than 1 in (1-2 cm) to approximately 3 ft (1 m). Their color depends on the type of algae that lives on them, although they can also be yellow, green, or brown. They exist by removing small particles, algae, and protozoans from the substrate by attaching themselves to wood or other reasonably strong substrate fragments. Freshwater environments are home to two common kinds of mollusks: gastropods, which include snails and limpets, and pelecypods. Snails can be herbivores or omnivores, consuming periphyton and occasionally detritus for food. The majority of freshwater bivalves are filter feeders, drawing fine bacteria, algae, and organic debris particles out of the water. As a result, they contribute significantly to the ecology by lowering turbidity, especially in lake environments. Mollusks are also eaten by fish, a few types of birds, crayfish, turtles, and mammals like otters and raccoons. Just around 4,000 of the roughly 40,000 species of crustaceans that are known to exist today reside in freshwater. Isopods, amphipods, decapods, and a few crabs are among the larger freshwater species.

Amphipods and isopods found in freshwater environments are omnivorous scavengers that live in pristine, cold waters. The decapods are herbivores, predators, and detritivores; some of them change their diets as they go through different life phases. Fish, birds, snakes, and mammals are just a few of the many predators that prey on these species. Some microcrustacean taxa are significant components of freshwater environment food webs. These comprise species belonging to the classes Branchiopoda, Maxillopoda, and Ostracoda. They live in a range of river habitats, mostly those that aren't subjected to strong currents, and eat organic matter as well as single-celled plants and animals. Although the macroinvertebrates in most rivers and streams are aquatic insects, the vast majority of species in this massive taxon are terrestrial. They spend their larval stages in the water, not their whole lives. For some species, the adult stage is quite short compared to the lengthy larval stage. With the exception of Antarctica, all continents are home to more than 2,000 species and more than 300 genera of mayflies(Vinson & Hawkins, 2003). Few species live in lakes and wetlands; the majority are obligatory river species. Fish rely heavily on mayfly species as a food supply in all freshwater aquatic environments.

Almost 10,000 species of caddisflies have been described, and there are likely many more of them in existence worldwide. Unlike many other orders of aquatic insects, they resemble moths and are all aquatic as larvae. The adults come out to mate and sometimes die in large groups. Caddisflies have developed a wide range of ecological specializations and are found in all freshwater habitats. Some graze on cows, plant material, predators, and some spin nets to gather fine particulate organic material (FPOM). They are widely distributed and a major source of food for fish. Stoneflies typically live on rocky or gravelly surfaces. This suggests headwater streams, which are actually the ideal habitat for most stoneflies due to their typically high oxygen content and cold temperatures.

The mid- and high-latitudes are typically home to the majority of the approximately 2,000 species. Adult life is short in comparison to that of the larval stage. There is only one stonefly species that lives its entire life in the water: the species found in Lake Tahoe. The whole range of feeding preferences is displayed by stoneflies: some are predators that eat midge and blackfly larvae, while others are detritivores whose life cycles are synchronized to take advantage of the season when leaves enter the stream. Certain species are detritivores at first, but as they mature, they become predators. Numerous fish species, birds, amphibians, and larger invertebrates like hellgrammites eat the larvae, eggs, and adults.

Figure 7.3 Midges

Source: https://commons.wikimedia.org/wiki/File:BitingMidge.jpg,*Creative Commons License.*

7.8 MIDGES

Midges (Figure 7.3) are widely dispersed and found in many types of aquatic environments. The majority are FPOM-feeding species. Predators make up a small percentage. Fish, amphibians, birds, and larger invertebrates all consume significant quantities of midges or chironomids. Crane flies also inhabit a wide range of environments, many of which are lotic. Though some of them are predators, they are mostly herbivores and detritivores. There are around 3,300 species in this huge order Hemiptera that live in or on water. The most noticeable is the water insect, sometimes known as the toe-biter (family Belostomatidae), which may cause painful bites. Few are big enough to consume fish, frogs, or water snakes in the same way as the water insect. Backswimmers, water crickets, and water striders are other members of this order. Dobsonflies, fishflies, and alderflies are some of the well-known aquatic insects in freshwater environments. One such species is the huge hellgrammite, or larval dobsonfly.

Most Megaloptera are detritivores in the early stages of their larval development, but they eventually transform into predators that eat a variety of invertebrates. Many traits of their larger counterparts are shared by alderflies and fishflies. Damselflies and dragonflies are found in freshwater areas. The majority of the population only see them in their adult, winged forms, which include numerous damselflies and dragonflies with striking colors. In their aquatic, larval, and adult stages, all are skillful predators that feed on aquatic invertebrates and small vertebrates before catching and consuming flying insects.

7.9 FISHES

Fishes are the predominant aquatic species. Relatively few vertebrates live their whole lives in water, even though other species inhabit freshwater habitats and are members of the aquatic food web. There are over 25,000 species of fish that have been formally described. Of them, around 41%, or over 10,000, are freshwater, and 1% spend their entire lives alternating between freshwater and the sea. Fishes have adapted to live in almost every type of aquatic habitat. They vary in size from < 0.5 inches, or 1 cm, to quite huge, with the whale shark having a maximum length of 12 meters or 39 ft. Certain fish are brightly colored; some fish have scales, while others do not. Certain animals consume plants (herbivores), while others consume invertebrates, debris, other fish, and occasionally, humans.

The majority of fishes are teleosts, or members of the division Teleostei, including the most common freshwater species. The Euteleostei, or real teleosts, are the significant subdivision in terms of both the total number of species and the number of significant freshwater species. The teleosts and other freshwater fishes are only distantly related. For instance, certain lampreys live in freshwaters and are members of the Petromyzontiformes order, which is distinguished by the absence of jaws. There are several cartilaginous fishes in freshwater, belonging to the Order Rajiformes, which is also referred to as skates. There are fish that are primitive, meaning they have survived in a form similar to that of their ancestors that lived hundreds of millions of years ago. These comprise the paddlefish and 24 species of economically valuable and heavily overfished sturgeons, as well as many species of lungfish in the orders Ceratodontiformes, Lepidosireniformes, and Acipenseriformes.

There are 23,600 species and 38 orders of teleosts, which are divided into 426 families. Among these, about 22,000 species are classified as real teleosts. The order Cypriniformes, which comprises minnows, is a vast group that includes the cyprinid family, which is the biggest family of freshwater fish(Mayden & Chen, 2010). However, there are no cyprinids in South America where characids appear to be equivalent to cyprinids. The Characiformes is a significant tropical order that contains piranhas that live in the Amazon basin and other places. The order of Siluriformes, which includes catfishes, is large and diversified.

The Protacanthopterygii, which comprises the order Salmoniformes, is another superorder of teleosts that includes several freshwater families. This order includes graylings, trout, salmon, chars, and whitefish. With over 12,000 species, the majority of which are marine, the superorder Percomorpha also includes several freshwater species. With 148 families and 9,300 species, one order in this category, Perciformes, is actually the largest order of vertebrates. It contains the Centrarchidae family of fish, which includes sunfish and crappies. The Tennessee River watershed in the United States is home to the perches and darters, which belong to the family Percidae. With almost 1,500 species, the huge family Cichlidae is important throughout the tropics, especially but not just in Africa. Cichlids have shown diversification in several African lakes, where the majority of the species are endemic. These lakes are home to a variety of cichlid species that differ in size, shape, color, and behavior.

7.10 AMPHIBIANS AND REPTILES

All amphibians require some type of freshwater habitat; many of them like lentic (still water) environments. Lotic (moving water) habitats are home to members of three orders: Anura, which includes frogs and toads; Gymnophiona, which includes caecilians; and Caudata, which includes salamanders and newts. Many frogs and toads spend their entire lives closely associated with freshwater bodies, breeding in streams or near their borders. Freshwater food webs involve a variety of reptiles, including members of the order Crocodilia, snakes (Squamata) belonging to various families, and freshwater turtles(Valencia-Aguilar et al., 2013). Alligators, crocodiles, and other related species are avid eaters that can consume invertebrates in addition to fish, particularly in the early stages. There are some alligators and crocodiles that reach huge sizes, and their effects on the fish population are probably significant. They live largely in lowland areas, slow-moving rivers and streams, lakes and ponds, and wetlands. While terrapins are often omnivorous, snakes are nearly exclusively carnivores, feeding mostly on fish but also on invertebrates.

7.11 BIRDS AND MAMMALS

Numerous birds inhabit marshes, rivers, lakes, and other bodies of water. They are part of the aquatic food chain and are frequently top predators. Although they are fish-eaters (piscivores), eagles, ospreys, kingfishers, and other birds are not restricted to water-logged environments, nor do they spend a lot of time in the water. Many birds that are not in any way aquatic are fed by the adult stages of aquatic bug larvae. However, a large number of bird species can be classified as really aquatic because they live a significant portion of their lives in the water or underwater. For example, cormorants and anhinga are comfortable hunting fish underwater as they are in the air. Certain aquatic birds exhibit a preference for fast-flowing water. A number of ducks search fast-moving waters for crustaceans. Gaviiformes live in rivers, lakes, and wetlands in the northern hemisphere's upper latitudes. They travel underwater with the help of their strong feet and eat a wide range of aquatic species, mostly fish, amphibians, and crabs. Podicipediformes are about 20 different species of grebe, including the New Zealand Dabchick. In both the Northern and Southern Hemispheres, they are widely dispersed. Pelecaniformes are seabirds, however, the common freshwater diving cormorant (*Phalacrocorax* spp.) also feeds water snakes in addition to fish. Ciconiiformes hunt in the shallow water of marshes, lakes, and slow-moving rivers for fish, crustaceans, amphibians, and other animals.

Gruiformes live mostly or only in freshwater habitats. There are over 130 species of rails, coots, crakes, and flufftails in the family Rallidae(Colyn et al., 2017). These birds usually live in marshes and around lakes and rivers in thick vegetation, where they consume food such as tiny fish, snails, insects, and seeds. The family Gruidae contains fifteen species of cranes, including the critically endangered Whooping Crane. Other aquatic or semi-aquatic Gruiformes members include the tropical Sungrebe and Finfoot, the Sunbittern, and the crane-like Limpkin. Terns, gulls, and shorebirds are some species of complex order Charadriiformes inhabit freshwaters, particularly wetlands, although many are restricted to the seacoast. For example, woodcocks excavate insects and worms in the bogs and swamps of the Northern Hemisphere. Although they are frequently seen near beaches, sandpipers can also be found near the edges of lakes, wetlands, and rivers. Numerous species of Charadriiformes can be found in salt lakes as well as coastal zones. This order also includes the terns (family Sternidae) and gulls (family Laridae), which are close relatives. Though they are frequently linked to marine settings, a large number of these birds also live in freshwater habitats, where some consume mostly fish and others mostly feed on insects. Because they are at the top of the aquatic food chain in lakes like Lake Ontario, gulls are more vulnerable to the impacts of bioaccumulated hazardous substances. The family Jacanidae, or jacanas, is well-represented in the Tropics. Their massive feet, which help to distribute their weight, give them the unusual ability to walk on lily pads and other floating vegetation. They prefer to live at the boundaries of wetlands and shallow lakes, where they can feed on insects and other invertebrates. Kingfishers are among the most noticeable species to live in many places of the world beside streams, rivers, and their banks. While certain kingfish species are not found in freshwater habitats, two families, the Cerylidae, which includes the Belted Kingfishers, and the Alcedinidae, which includes the River Kingfishers, are the majority of kingfishers live in Eurasia and Australasia. The family includes all American Kingfishers, notably the Belted Kingfisher.

While many mammals are linked to aquatic food webs and use riparian zones, very few live in the water. The Australian platypus, beavers (family Castoridae), hippopotamus (family Hippopotamidae), river dolphins (family Platanistidae), otters (family Mustelidae, subfamily Lutrinae), and a few other mustelids are among the mammals that do spend a large portion of their life in freshwater.

7.12 THE DYNAMICS OF STREAMS

The freshwater aquatic ecosystems that are dynamic are those found in rivers and streams. River channels, riparian regions, floodplains, riffles and pools, and many other physical characteristics related to river systems are shaped by the movement of water. Moreover, the flow volume varies dynamically. A fish or other aquatic macroinvertebrate faces both opportunity and risk from the dynamic movement of the water. Velocity rises and falls with changes in the rate of flow. If there is too much velocity, some species run the risk of being carried away; if there is too little, those that rely on the river to carry food will go starving. Understanding the lotic environment, or flowing-water environment, is essential to comprehending how species have adapted to live in it. The amount of water that flows through a certain spot on a stream or river in a specific amount of time is known as the discharge, also known as streamflow. The physical morphology of natural stream channels varies widely. In some areas they are wider, in others narrower; in some places, they are deeper, in others shallower. This makes it difficult to measure discharge accurately. For instance, velocity is slower toward the bottom and edges of the channel and greatest near its middle.

Changes in discharge cause variations in velocity over time at a single point on a stream. In most natural stream channels, when the discharge later decreases, breadth, depth, and velocity all decrease as well. This is because more water is supplied into a stream, for instance, as a result of an unexpected downpour, which raises width, depth, and velocity. As a result, river species' access to habitat varies in quantity and quality. Different places along the stream may experience varying velocities at constant discharge. Velocity is determined by three factors: viscosity, slope, and abnormalities in the channel. The ratio of the vertical drop to the horizontal distance, or the steepness of the channel, determines the slope of any stream or river reach. It can be expressed either as an angle or as a percentage. A river's slope can occasionally be expressed in feet per mile, which is the amount of elevation loss in feet each river's horizontal mile. An extremely steep section of the Upper Youghiogheny River descends about 200 feet each mile. The Amazon, on the other hand, has widespread flooding during the rainy season because, once it departs the Andes, it decreases only 0.053 feet/mi over 2,500(Marengo et al., 2012). The force of gravity, which acts perpendicular to the Earth surface, provides the energy that water in a river can use. As a result, the acceleration caused by gravity pull increases with slope steepness. Steep mountain streams resist freezing because turbulence in natural rivers converts most of the energy of flowing water into heat. Any feature in a river channel that hinders flow and creates turbulence is referred to as a channel irregularity. Here, variations in channel roughness, width and depth are taken into account. Channel meanderings can create turbulence and dissipate energy. Velocity is influenced by sediment load, which is the quantity of suspended silt, sand, gravel, and other elements that the stream is carrying. The velocity decreases as the sediment load increases. One factor that affects velocity is viscosity. Viscosity in water changes with temperature, with colder water being more flow-resistant. Viscosity has less of an impact on velocity than slope, roughness, and sediment load.

The mobility of the water and the physical conditions that follow are two factors that contribute to the dynamic and varied conditions under which river plants and animals have evolved and survived. The enormous diversity of ecosystems produced by flowing water is reflected in the physical structure of rivers.

7.13 THE AMAZON RIVERS

Among rivers, the Amazon is large. More water is sent to the sea by it annually than by the combined volume of the six largest rivers, the Congo, Yangtze, Orinoco, Brahmaputra, Yenisei, and Rio de la Plata. Though each of these is a massive river, the Amazon exceeds them all. It has 3,000 different species of fish and is the largest watershed. It releases so much silt and freshwater into the Atlantic that it changes the ocean salinity and color for 200 miles (321 kilometers) offshore. In addition, the river ecology is changing quickly. According to estimates, the Amazon watershed is between 2.4 and 2.9 million mi Km^2. It encompasses a large portion of Brazil as well as portions of Bolivia, Peru, Ecuador, Colombia, and Venezuela. Because the water drops very little on its journey to the sea once the headwater rivers leave the Andes, the watershed has a different form. The Amazon has a low slope of less than an inch per mile since, aside from the Andes, a large portion of the basin and the middle basin are relatively low. One explanation for this is that the Amazon is a river with a reversed flow. It used to run from east to west and empty into the Pacific. Subsequently, the river was obstructed to the west and could only flow eastward upon the emergence of the relatively young Andes. Its vastness is matched by the extraordinary diversity of its landscapes, terrestrial ecoregions, and landforms, ranging from millions of acres of marshes barely above sea level to the Andes, which rise to a height of 22,000 feet in the west. The distribution of land cover is as follows: 73 percent is forest; 10 percent is grassland, savanna, and shrubland; 8 percent is wetlands; 14 percent is agricultural; 6 percent is dryland region; and less than 1 percent is urban/industrial area.

The river extends 3,900 river miles (6,375 river kilometers) from the sea in the Peruvian Andes glaciers. It was only recently that a global positioning system was used to locate that source, which is a thin sheet of clear water pouring across a rock face on the slopes of Nevado Mismi, an 18,362 ft (5,597 m) mountain not far from Lake Titicaca. The Rio Apurimac has lately been recognized as the source headwater, but the Rio Maranon was long thought to be the source. The Ucayali, which combines with the Maranon to form the Amazon, is the source of the Apurimac(Mayorga et al., 2005). As far down as Manaus, where the Rio Negro enters from the north, the mainstem river, which was formed by the Ucayali and the Maranon, was once known as the Solimoes. It wasn't until then that it was regarded as the Amazon. The Amazon is joined downstream by the Jurua and Purus from the south and the Japura from the north, at the confluence of the Ucayali and the Maranon. After that, the Rio Negro enters from the north, and the Madeira, another significant tributary, enters from the south a few hundred kilometers downstream. The Rio Negro has a mean average annual discharge of more than 3.5 million ft3/sec, which is greater than any other river in the world that saves the Amazon. These major tributaries are themselves very huge rivers. Although the amount of rain varies from place to place, the average rainfall is roughly 79 in (2,000 mm). There are areas of the basin that receive up to 26 feet (8,000 mm) of precipitation annually, especially the eastern slopes of the Andes, where orographic precipitation happens. Despite the high rates of evapotranspiration brought on by the tropical temperatures and strong sunlight, there is still a significant amount of runoff from this rainfall. Given that the basin crosses the Equator, temperatures do not fluctuate much throughout. Until the far western borders of the Andes, there is minimal variation in altitude.

A location with minimal topographic relief and a river with an exceptionally low channel slope result in an annual inundation of the floodplain forests for weeks at a time. Varzea refers to the floodplains along the Amazon and its tributaries, but it can also refer to any of the forested floodplains that experience frequent flooding(Junk et al., 2011). Every year, the waters of the increasing rivers that drain the Andes and their foothills bring in organic matter, nutrients, and fine sediments that sustain these ecosystems. Fish and other aquatic life are nourished by these flooded forests in turn. The flooded terrain is known as igapo on black-water and clear-water streams, like the Rio Negro and others draining the soils north and south of the Amazon. Here, flooding results in the emergence of a distinct terrestrial and aquatic community compared to the várzea forest. However, during the high tide period in both situations, the floodplain is submerged, sometimes to remarkable depths. Many lakes are left behind by the retreating water during low tide. Then, when evaporation shrinks them, plants develop and decay and oxygen conditions change, these go through ecological changes. When the river reclaims these lakes at high tide, the fish immediately go in search of greener pastures in the flooded woodlands. The Neotropical biogeographic region, which includes the Amazon Forest, is incredibly diverse biologically. This also applies to the river system, which is not surprising given how deeply entwined the river and forest are, making it occasionally difficult to discern where they part. The floodplain has seen 30% of the water that the Amazon releases into the sea.

7.14 HABITATS

The many river habitats including floodplain lakes, which are a seasonal feature of the river, offer a range of circumstances and complications for photosynthetic species. The plants that arise from this condition are incredibly diverse. The rivers of Amazonia are generally home to four different types of vegetation: woody plants, aquatic herbaceous plants, terrestrial herbaceous plants, and algae. The combination of plant species present at any given location and time on the river, including its floodplain and the lakes within it, is dictated by gradients of inundation and human activities. In this area, these gradients appear slightly as zones of height, succession, and substrate stability as determined by erosion and sedimentation. Furthermore, distinct plant relationships arise from the Amazon basin's multitude of diverse ecosystems. Despite warm temperatures and abundant sunlight, phytoplankton biomass in the rivers and streams of the Amazon is comparatively low, despite a great species variety. Due to a shortage of nutrients, productivity is generally poor in clear and black-water streams. The amount of light in streams in low-order forests that are fully covered by trees restricts the growth of phytoplankton. Though there are enough nutrients in whitewater streams, turbidity prevents light from penetrating and thereby inhibits the growth of phytoplankton.

Floating mats, sometimes known as floating meadows, are abundant in whitewater rivers, such as the Amazon, and can support entire ecosystems of plants and animals. In the floodplains, grasses spread swiftly on exposed sediments during low tide. These colonies grow floating stems with hanging roots when high tide returns, effectively lifting off with the rising water. The underside of the floating mats provides a habitat for the vast growth of periphyton, which in turn supports a complex web of fish and zooplankton. The floating mats are heavily populated by insects and their predators. When microcrustaceans or insect larvae come into contact with the underwater carnivorous bladderworts growing on the floating mats, they burst, swallowing and digesting their prey. Deposition and erosional features are rapidly changing. Sand-filled soils might be prone to flooding throughout the year, only drying out during the lowest point of the low tide phase. When condition favors, annual terrestrial plants in this area proliferate and spread quickly. Certain species have the ability to float freely and integrate into the plant matrix of floating mats when the water level increases. Stronger currents and wave action can affect steeper riverbanks. Water paspalum, aleman grass, and Mexican crowngrass are a few of the more uncommon species that grow on such banks.

Low-lying flats and lake beds in floodplains acquire fine deposits during their extended period of flooding and are not exposed to strong currents or erosion other than that caused by waves. They sustain a relatively low diversity of plants during their brief dry period, including sedges like sedges and grasses like West Indian marsh grass. More aquatic species, such as some kinds of wild rice and free-floating plants, replace the main terrestrial species as the water level rises. Large concentrations of phytoplankton can be found in floodplain backwater areas and lakes during low tide periods. Large populations of diatoms and cyanobacteria feed off the reduced turbidity and lack of turbulence in certain lakes, causing them to become eutrophic(Reynolds, 1994). Similar to low-lying flats, low-lying swales retain enough soil moisture to sustain certain aquatic free-floating plants even during the low-tide period. Even with the prolonged flooding, trees and bushes are still supported on the floodplains. The diversity of the floodplain woods is a reflection of the dynamic and complicated landforms. While natural levees where the river first deposits its maximum sediment load as it enters the floodplain and the transitional zones to the terra firme are dry enough for woody vegetation to grow, the floodplain lakes and swales are moist for it. There are trees in close proximity that are in various phases of succession due to the dynamic terrain generated by the rivers as they meander back and forth. The Central Amazon Varzea is characterized by a mosaic of plant communities with a majority of evergreen wet tropical forest. The composition of these communities is influenced by various factors such as soil conditions, frequency and duration of flooding, and seral stage. *Salix martiana*, a willow, and Iporuru are examples of early seral shrubs. As the water rises, this willow and many others in the flooded forest form adventitious roots to carry oxygen from the air to the sections of the plant that are submerged.

Figure 7.4 *Cecropia latiloba*

Source: https://commons.wikimedia.org/wiki/File:Cecropia_latiloba_%2817224873741%29.jpg,*Creative Commons License.*

7.15 SUCCESSION

Later successional woody plants include fish-dispersed successional trees, a spiny palm whose seeds can withstand 300 days in the water, and *Cecropia latiloba*, (Figure 7.4) a tree that effectively colonizes the nutrient-rich floodplains of whitewater rivers due to its extreme tolerance to flooding, submergence, intense sunlight, and sediment deposition. It may grow vertically fast and its seeds are resistant to submersion. Species from the genera Ceiba or Chorisia, Spondias, and Virola, a nutmeg family member, are examples of late successional trees. Compared to the várzea forest submerged by whitewater streams, igapo forests, which are inundated by nutrient-poor waters, are home to fewer species. The timing of the flood affects several tree life cycle events in flooded forests, such as fruiting, flowering, seed production, dispersal, and germination. Mayflies, stoneflies, and caddisflies were the most common species. The lowland river benthic invertebrates have not received much attention, apparently because most species find the shifting sand substrate unsuitable. This sandy substrate stretches 2,485 river miles upstream from the Amazon mouth. However, a wide range of insects can be found in the low-order headwater streams of the black and clear-water tributaries. The most varied insects in these streams are caddisflies, black flies, and midge flies. Mollusks are virtually missing, as well as other cold-water species like stoneflies, which are thought to be restricted by the low concentrations of dissolved minerals. Fungi that feed on allochthonous debris from the nearby forests, which are then eaten by small chironomids and then by a variety of carnivores, constitute an important food chain in these streams. Large lowland rivers may not provide adequate circumstances for benthic invertebrates; however, floodplain lakes can occasionally offer favorable conditions for relatively high concentrations of invertebrates. However, many of these lakes experience low oxygen levels when the floods pass, especially at the bottom, which lowers the number of invertebrates(Wantzen et al., 2016).

There are claims that the largest diversity of aquatic invertebrates can be found in and around the floating mats that are present in the big rivers and flooded várzea. The periphyton that grows amid the roots of the floating macrophytes in them serves as the foundation for a flourishing invertebrate population known as the perizoon. Along with submerged or emergent macrophytes, the perizoon colonizes floodplains as they are flooded. Water mites, copepods, ostracods, water fleas, and fly larvae are among the many organisms that can be found in whitewater streams, especially in mats. Larger invertebrate predators and numerous fish species eat them. Additionally, the mats are home to snails belonging to the genus Biomphalaria, which is responsible for schistosomiasis, a disease that is common in tropical regions. Due to their abundance, a variety of invertebrates found on floodplains play a significant role in the river food chain. Blackwater and clearwater streams are home to an uneven number of freshwater shrimps. On the other hand, the Amazon River prawn lives in whitewater streams and is a valuable food source for numerous fish species. Large and abundant, apple snails play a significant role in the feed of caimans, certain birds, and a number of fish species.

The snail-kite feeds almost solely on snails, and its beak is designed to fit within their shells. Another significant invertebrate link in the food web of the whitewater rivers and floodplain is a type of mayfly due to its large population. It is skilled at digging into dead wood during its larval stage to speed up its breakdown. Estimates of the number of fish species in the Amazon basin range from 2,000 to 4,000. Of the species in the basin, 40 percent belong to the characid (order Cypriniformes) group, and 25 percent to the catfish (order Siluriformes) group(Reis et al., 2016). There are also 11 other orders of fishes, which constitute 21 families, some of which have numerous species. Fish diversity in the Amazon basin is a result of the enormous diversity of habitats. There is not much of a difference between the species diversity of blackwater and whitewater rivers. The species diversity in both systems is greater than that of streams of the same order found in temperate climates. Fish dispersal is determined by physical factors like oxygen content, water chemistry, and the existence or lack of lotic conditions. The high dissolved mineral levels in the whitewater streams are comparatively intolerant to certain organisms found in blackwater rivers. Since the whitewater Amazon mainstem would provide a barrier to propagation into adjacent black-water streams for such species, it is believed that this would result in a higher degree of endemism within specific tributaries. Nonetheless, a large number of species that are not subject to such restrictions use the Amazon mainstem as a route for dispersal, either by actively migrating or by using the currents to spread their young. Some fish disperse into the várzea or igapo at high water and stay in floodplain lakes during low water, avoiding the Amazon mainstem and its huge, lotic tributaries entirely. These animals most likely developed morphologically or behaviorally to cope with low oxygen environments, which are typical in Lagos during the low water season. For instance, the Black Acara can endure low oxygen levels by drawing air into its stomach. In order to effectively absorb oxygen, other taxa have developed unique adaptations for their mouths, gill cavities, swim bladders, and intestines.

Others, especially tiny fish, live in a shallow surface area of Lagos that receives oxygen from the surrounding air. Others still move away from macrophytes and toward open water at night when periphytic production stops oxygenating the water; during the day, they move back to take advantage of the oxygen supplied by the macrophytes themselves, many of which have physiological adaptations that allow them to absorb atmospheric oxygen into their submerged portions, and by the periphyton that lives on them. Small to medium-sized fish with a range of habitats, the characids most likely attain their maximum population in the Amazon basin. The Amazon basin is home to 15 genera of characids, which contain several species, including tetras. It also includes the piranhas, of which the Amazon is home to 28 species, which are the well-known fish in the region. Their large population size has a significant impact and is a major component of the feed of birds, dolphins, and people in the Amazon. The majority of the many species of turtles found in Amazonia are aquatic. The largest of them all is the Amazonian River turtle, (Figure 7.5) which can grow to a length of 25 to 30 inches. Perhaps the most well-known is the unusual-looking matamata, which fills a similar ecological niche to that of the North American alligator snapping turtle. Both are opportunistic piscivores that hunt on the bottom of primarily lentic waters, using camouflage to wait for their prey. Though their methods of catching fish are slightly different, the matamata catches fish by striking out with its jaws open and sucking in water, thereby drawing the fish into its mouth. The Amazon floodplains are home to a large number of different reptiles and amphibians, but very few, if any, of them are aquatic species. The anaconda is a popular reptile in the system; however, it may not be as huge as some suggest. Though by no means small, its maximum length is approximately 40 feet, which is less than the occasionally stated 65 feet. Because the majority of its prey is aquatic, it frequently visits rivers. Fish, turtles, caimans, and capybaras, large rodents that inhabit floodplains, are their primary food sources.

Figure 7.5 Amazonian River Turtle

Source: https://en.wikipedia.org/wiki/Yellow-spotted_river_turtle#/media/File:Podocnemis_unifilis_in_Zoo_Krefeld_(2).JPG, *Creative Commons License.*

7.16 AMAZON WETLANDS

The majority of industrialized nations and many developing nations concurred at a conference held at Ramsar, Iran, in 1971 that wetlands are important environmental resources that need to be preserved. The ideas of the Ramsar Convention, as it is now called, were accepted by 153 nations by 2006, and many nations including the US now have national laws in place to conserve wetlands(Griffin & Ali, 2014). The belief held by the majority of people for hundreds of years that wetlands were, at best, wild environments, uninhabited impediments to human progress, and, at worst, waste places harboring disease, has completely changed with the global consensus that wetlands deserve conservation. Despite the good intentions of governments, many people in society still have outdated beliefs about wetlands. In addition to removing sediment and nutrients from streams, which could pollute them, wetlands also serve as a habitat for a variety of species, and offer recreational opportunities. They also control and prevent downstream flooding. Wetlands are distinguished by three features: water, or more precisely, the level of water relative to the land surface; soils that exhibit the revealing marks of being saturated with water all or most of the time (hydric soils); and the presence of water-loving (hydrophilic) plants. U.S. government agencies utilize these three elements to identify wetlands on the ground. Practically speaking, identifying wetlands is crucial because, should an area be designated as one, there may be limitations on how the property may be used. The land's worth may be significantly impacted by these limitations. Environmental planners, land managers, biologists, and ecologists, as well as wetlands scientists, can operate more easily due to the development of wetlands classification systems. The factors that determine the quality and trophic status of water include hydrology, which includes hydrodynamics and hydroperiod, water source, freshwater versus saltwater, position in the landscape, and vegetation, which combines the effects of climate and biogeography with other factors. Wetlands are exactly what their name suggests: wet plains. Which plants and animals can be found in a wetland depends on its level of moisture and duration. A wetland's hydroperiod, or the signature of water on that specific piece of land, is the pattern of rises and falls in water level throughout time in that wetland. The variations in the water level can be attributed to several factors such as fluctuations in groundwater levels, surface or subsurface lateral flows caused by rising and falling water levels of nearby lakes or rivers, and surface and near-surface runoff of precipitation.

There are instances where the duration of surface inundation is used to categorize hydroperiods. In wetlands that are not affected by ocean tides, known as non-tidal wetlands, there are various types of hydroperiods that occur there. These include permanent or year-round flooding, intermittent or year-round flooding except in extremely dry years, seasonal or yearly flooding during the growing season, saturated (rarely standing water present but annual soil saturation during the growing season), and temporary or brief flooding during the growing season followed by a period of time when the soil is not saturated. These hydroperiods can be classified as subtidal or flooded with tidal water even at low tide, irregularly exposed or the substrate is exposed by the tides less frequently than daily, for example only during unusually low tides, regularly flooded or flooded then exposed at least once daily, and irregularly flooded in tidal wetlands, including freshwater tidal wetlands or flooded less often than daily. The hydroperiod in nontidal wetlands is mostly determined by evapotranspiration and precipitation. In America, vernal pool wetlands experience low evapotranspiration rates in the winter and spring, when water is above ground. The water table lowers and evapotranspiration rates rise during the warmer months(Manning et al., 2013). The ups and downs of their rivers are linked to the hydroperiods of riparian wetlands, or wetlands next to rivers. Every year, for many months, the Amazon River floods the Brazilian Amazonian forests. The water level rises sharply from below the surface of the ground to many feet above. The creation of the distinctive soils of a wetland requires at least one week of flooding throughout the growth season. The water budget, or the inputs and outputs of water to a wetland, can be used to examine changes in water level. Potential water inputs or sources include surface flow into the wetland, direct precipitation into the wetland, and groundwater discharge into the wetland. These sources may vary depending on the type of wetland. Evapotranspiration, surface runoff from the wetland, and percolation back into groundwater are examples of potential water outputs. Nutrients, silt, organic matter, dissolved solids, and organisms may arrive with water. For instance, flooding rivers frequently provide significant amounts of silt, fertilizers, and organic elements to wetlands in floodplains. The materials that remain after the wetland water evaporates entirely through evapotranspiration include nutrients, silt, organic matter, and dissolved solids. A salt buildup from this might eventually transform freshwater into a saltwater marsh. Over time, the accumulation of sediment and organic material can transform a wetland into a different type, possibly extending the period of saturation as soil pore spaces get blocked with fine particles and organic material.

7.17 WATER TABLE

The word water table refers to the level of water that controls the amount of moisture in a wetland. The top of the saturated soil subsurface zone, or the zone where water instead of air fills the pore spaces between soil particles, is known as the water table. The water table in non-wetland regions can be found a few yards to hundreds of feet below the surface of the earth. The water table is typically near or even above the ground surface in wetland environments, which causes water to come to the surface. The exceptions are wetlands known as ombrotrophic wetlands, which are only supplied by local precipitation such as rain and snow. For these types of wetlands to develop at all, the soils on which they are found must be highly impermeable, meaning that water cannot easily move through them; instead, it gathers in surface depressions where a wetland eventually emerges. The trophic state of a wetland is directly correlated with its water source. Although there are some exceptions, such as the flooded forests of the nutrient-poor black-water and clear-water tributaries of the Amazon, wetlands connected with river systems and wetlands fed by groundwater often have higher levels of important plant nutrients than those fed simply by rainfall.

Hydric soils are the type of soil found in wetlands(Moorhead & Cook, 1992). Hydric soils are those that developed under conditions of saturation, and flooding, long enough throughout the growing season to establish anaerobic conditions. These anaerobic conditions are defined as low oxygen levels brought on by water saturation. Because they are rarely saturated, the majority of non-hydric soils are aerobic. In these kinds of soils, plant roots make use of the easily accessible oxygen to sustain their growth. Because there is oxygen in the soil to support the decomposer activity, organic material, such as dead plant roots, leaves, stems, and soil species, decomposes quickly in aerobic soils. Organic material decomposes slowly in anaerobic soils; in certain circumstances, decomposition may proceed so slowly that the annual addition of organic material exceeds the annual removal of organic material through decomposition. Nearly all of the upper soil in peat wetlands is organic, and even after centuries, certain plant fragments are still immediately distinguishable. Soils found in wetlands can be categorized as either mineral or organic. More than a third of the material in organic soils is organic in nature. They have a higher capacity to hold water, like a sponge, than mineral soils. Additionally, water easily flows through them due to their high porosity. Consequently, the formation of wetlands with organic soils is contingent upon the presence of a slightly impermeable layer beneath the hydric soils to prevent water from seeping into the groundwater system. Organic soils are typically low in nutrients and acidic. Because of frequent or continuous saturation and a low rate of decomposition, often brought on by cold temperatures and brief growing seasons, they form in anaerobic conditions.

Chapter 8

INTRODUCTION TO DESERT BIOMES

Contents

8.1 INTRODUCTION

A warmer or drier environment leads to the formation of deserts. Because more water will evaporate from the oceans as the temperature rises, the two are not necessarily the same. There's going to be more rain and clouds. The rate of evaporation will decrease with a drop in temperature. Rainfall and clouds will decrease. During the last Ice Age, global temperatures were higher than they are now. Therefore, it would seem that deserts should get smaller as the temperature rises, but this isn't always the case. The rate of evaporation rises with warmer weather, but if it does so at a faster pace than the rainfall, the ground will get drier even though the amount of rainfall has increased. Additionally, the rate at which water vapor condenses is lowered at higher temperatures. Because the air can contain more moisture in the form of water vapor as it gets warmer, less cloud cover and less rainfall occur even as the air moisture content rises. Although they may form if average temperatures rise, deserts are more likely to form in a cooler climate. Hydrogen bonds, which are formed between the hydrogen (H) atoms of one molecule and the oxygen (O) atoms of two neighboring molecules, hold groups of water molecules together to form liquid water (H_2O). Individual molecules vibrate, and molecules move and slide past one another. The molecules gain energy as the temperature rises. The molecules move quicker and vibrate more intensely. An increasing number of molecules absorb enough energy as the temperature rises to separate into individual water vapor molecules and escape from the air(Garrett et al., 2006).

8.2 SCIENCE OF WATER

Water molecules that enter and exit the liquid at the same time as water vapor molecules impact the liquid surface. Water evaporates when more molecules exit the liquid than enter it; this process happens more quickly at higher temperatures because the molecules lose energy as they evaporate or vaporize. Due to the water molecules that have entered the air during evaporation, the air pressure rises and continues to climb as more and more water molecules escape into the atmosphere. Since it represents the percentage of total air pressure that is attributable to water vapor, the increase is known as the vapor pressure. But more molecules are forced back into the liquid when the vapor pressure is raised. When the quantity of water molecules entering and exiting the liquid equals one, a tipping point is eventually attained. To put it another way, condensation and evaporation are balanced. At this point, the mixture of air and water vapor is saturated and the vapor pressure has achieved the saturation vapor pressure. The rate of evaporation increases with rising air and water temperatures. As more water is introduced into the atmosphere, the saturation vapor pressure rises.

Warm air is able to contain more water vapor than cold air because it requires a high vapor pressure before condensation equals evaporation. One pound of dry air can store 0.27 ounces of water vapor at freezing temperature and sea level pressure. One pound of air can carry two ounces of water vapor at 86°F (30°C), and 31.5 ounces at 104°F (40°C). In contrast, one pound of dry air can only store 0.008 ounces of water vapor at a temperature of –40°F (–40°C). Humidity is the measure of how much water vapor is in the air. Relative humidity (RH) is the most commonly used method. This represents the water vapor content in percentage relative to the air saturation point. The saturation vapor pressure increases with temperature, whereas the relative humidity decreases. The air is essentially drier due to the higher saturation vapor pressure even if no moisture has been added or removed from it.

8.3 PLANTS

Plants wilt as the land dries. If there is significant rainfall, they will eventually recover; but, if there is no more water available, there is no chance of recovery. The plants wilt and die. The root hairs that held soil particles together are slowly destroyed. The structure of the soil is lost. Clay soils have deep, narrow cracks and become as hard as concrete as they dry up and split(Tang et al., 2021). Sandy soils produce fine grains, while silty soils produce dust. The wind mixes grains and dust. When they land on adjacent land, they cover or even bury nearby plants. That is the way the desert grows. Libya is a large country. Although it rarely rains due to its location inside the Sahara, coastal cities like Benghazi and Tripoli do experience some winter rainfall. But the weather wasn't always so arid. There are caves in the Tibesti Mountains that have wall paintings dating back between 7,000 and 8,000 years that are located on the boundary between southeast Libya and Chad. Deserts arise and disappear due to changes in climate, and this is not limited to the Sahara. A climatic optimum occurs when warm, humid weather persists for an extended length of time and impacts a wide area. The climatic optimum that the hunters of the Tibesti Mountains experienced was most likely the warmest and longest since the end of the most recent Ice Age. It had an impact on the entire world and lasted between 10,000 and 4,000 years ago. Rainfall in the Middle East between 6,000 and 4,400 years ago was so intense that it caused floods in several ancient cities, including Ur and Nineveh in modern-day Iraq. Around 4,000 years ago, people were producing melons, dates, wheat, and barley in what is now the Thar Desert, on the boundary between India and Pakistan. At that time, the annual rainfall there was 16–32 inches, and Australia had a wetter climate than it does now. The Thar Desert currently receives five to ten inches of precipitation annually. A civilization based on the towns of Mohenjo Daro and Harappa flourished in the Indus Valley to the west of what is now the Thar Desert between 4,500 and 3,700 years ago, as that period of maximum productivity was coming to an end(Clarke et al., 2016).

8.4 CLIMATE AND DESERTS

Climates are ever-changing. Deserts grow and recede as a result of these changes, and civilizations have risen, died, and collapsed in response. The largest of the subtropical deserts, the Sahara, spans more than 3.5 million square miles. There are mountainous areas in the Sahara. Standing at 10,712 feet high, Mount Tousside is an extinct volcano located in the Tibesti Mountains. Low-lying basins can be found elsewhere, although the majority of the desert is situated on a plateau 1,300–1,600 feet above sea level. The old divisions of the desert include the Atlantic, northern, central, southern, and eastern regions. The Libyan and Nubian Deserts are further divisions of the vast Eastern Sahara. The Arabian Desert includes the desert in Egypt to the east of the Nile River.

Figure 8.1 Sahara Desert

Source: https://upload.wikimedia.org/wikipedia/commons/c/cb/Libya_4985_Tadrart_Acacus_Luca_Galuzzi_2007.jpg,*Creative Commons License.*

The Sahara (Figure 8.1) extends eastward to become the Arabian Desert, which occupies almost the whole Arabian Peninsula, as well as the Syrian Desert, which encompasses a large portion of the Middle East. The Tropic of Cancer in the Northern Hemisphere and Tropic of Capricorn in the Southern Hemisphere are the centers of the subtropical deserts. The subtropical deserts of the south cover a smaller area than those of the north because of the arrangement of the continents on Earth's surface, which results in less land in the southern hemisphere than in the northern. A series of mountains that runs parallel to the coast of the Red Sea and rises to elevations of over 9,000 feet above sea level extends down the western edge of Arabia. The highland refers to the region in the center of Arabia that is east of the mountains(Alsherif & Fadl, 2016).

Figure 8.2 Syrian Desert

Source: https://upload.wikimedia.org/wikipedia/commons/9/97/Syrian_Desert.png,*Creative Commons License.*

The Syrian Desert (Figure 8.2) stretches from Nafud northward to southeastern Syria, eastern Jordan, and western Iraq. It is a hybrid of grassland and desert. The Thar, also known as the Great Indian Desert, stretches over the Arabian Sea to the east and includes a portion of eastern Pakistan as well as about half of the Indian state of Rajasthan. The Thar, which translates to sandy waste, is a sandy desert that is roughly 77,000 square miles in size. The subtropical deserts of the Southern Hemisphere are the Australian deserts and the Kalahari Desert in southern Africa. The Australian deserts are approximately 1.3 million square miles, while the Kalahari is roughly 275,000 square miles.

8.5 DESERT CLIMATE

Although the eastern portion of the desert only receives roughly five inches of rain annually, the annual rainfall ranges from 10 inches in the south to

25 inches in the north. There is less than 10 inches of rain in the Australian deserts annually. The rate of evaporation in the Australian and Kalahari deserts is sufficiently high that the earth remains dry for the most of the year. The subtropical deserts are all quite hot during the day and can be cold at night. For instance, summertime averages at Timbuktu, Mali, in the southern Sahara reach 110°F (43°C), but wintertime lows have been recorded to approach freezing. Even hotter is Salah, Algeria, where July highs average 113°F (45°C) and can reach 122°F (50°C). However, in the winter, nighttime lows in Salah can drop below freezing. In many places of Syria, frost is a common winter weather condition. The Mongolian capital, Ulan Bator, has experienced below-freezing temperatures even during the summer. Situated in the steppe grasslands of central Asia, Ulan Bator is in proximity to the northern border of the Gobi Desert.

The eastern Gobi mining town of Dalanzadgad has no recorded weather records, while Jiayuguan, on the southern edge of the desert, has dry weather. Between October and February, there is very little rain, and between March and September, there is an average of 2.7–3.4 inches of rain. The average daytime temperature in January is 27–30°F, with evening lows of roughly 3–7°F. The winters are cold and the summers are pleasant but not too hot. In Jiayuguan, July is the warmest month with daytime highs of 83–86°F. The average temperature of Hohhot, a town near the eastern edge of the Gobi, is 9.1°F in January and 72.7°F in July. The Gobi is surrounded by mountains and is roughly 3,000 feet above sea level in the east and 5,000 feet in the west. Its surface is primarily made up of gravel and bare rock, with some areas having sand dunes. Approximately 500,000 square miles make up the entire area. A portion of it is in the inner Mongolia Autonomous Region and part is in southern Mongolia. Despite receiving barely one to two inches of rain annually, grass, thorn bushes, and other shrubs cover almost three-quarters of the whole Gobi.

8.6 GOBI DESERT

The Gobi is both a desert and a semidesert. It is the kind of desert that can be found outside of the Tropics in the deep interior of continents. The majority of the Tarim Basin, a low-lying region that borders the Tarim River in the Xinjiang Uygur Autonomous Region of China, is covered by the Taklamakan, a considerably dry desert to the west of the Gobi (Figure 8.3).

It is a sand desert that spans a region of over 105,000 square miles and features massive dunes that can reach heights of up to 300 feet. The weather is arid. The average annual rainfall in the western portion of the desert is 1.5 inches, whereas the average in the eastern region is barely 0.4 inches. Sandstorms are frequent and continue for many days. July temperature is 77°F (25°C), while it can occasionally reach 100°F (38°C). During winter, the average temperature is between 14 and 16°F, with occasional lows of –4°F.

Figure 8.3 Gobi Desert

Source: https://upload.wikimedia.org/wikipedia/commons/3/35/Gobi_Desert_dunes.jpg,*Creative Commons License.*

8.7 OTHER DESERTS

The Gobi and Taklamakan have an equivalent in the Southern Hemisphere, called Patagonia(Zech et al., 2022). In Argentina, the eastern Andes region and the southern Colorado River region, located around 39°S, are covered by this desert, which is located in the interior of a continent. It covers over 300,000 square miles in total. Nowhere in Patagonia receives as much rain as 10 inches, while the central region receives no more than five inches annually, despite having a wet climate than the Gobi and Taklamakan. In central Patagonia, average summer temperatures range from 45°F to 78°F in January; however, summertime highs of 99°F and wintertime lows of 3°F have also been recorded. Because weather systems from the west lose moisture when they traverse the Andes, Patagonia is dry. The desert is located under the rain shadow. The same can be noticed in the case of the Mojave Desert in North America. Stretching 15,000 square miles to the east and south of the Sierra Nevada, it includes this region. Despite the low annual rainfall of less than five inches, there are forests in the highlands and areas of the desert where cattle graze. Summertime highs frequently surpass 100°F. Wintertime highs are often around 55°F, but lows are far below freezing at night.

Because it is located on the eastern half of the continent, Patagonia is unique. Because weather systems approach it from the west, it has a dry climate. As it rises to pass the Andes, mild, humid air from the ocean evaporates on the western side of the mountains. Only very little rain can be delivered by the air by the time it reaches Patagonia. Subtropical deserts typically arise on the western margins of continents. This is due to the fact that the dominant winds there are directed toward the equator, bringing with them cold, dense air that stays near the ocean surface and keeps wet air from rising and creating clouds. Throughout the day, the sea stays cold as the land quickly warms. Cooler air from the sea replaces the warmer air that rises over the land. However, the air that is approaching must traverse cool ocean currents that run parallel to western coasts. The air tends to sink rather than rise as a result of the cold water's ability to reduce temperature. West Coast deserts are a result of these climate characteristics. This is how the air that has traversed the cool California Current produces the North American deserts. The Sonoran Desert, commonly known as the Yuma Desert and the Desierto de Altar, includes the Colorado Desert. While the average elevation in the Sonoran Desert is 1,000 feet, the bed of the brackish Salton Sea Lake in the Colorado Desert is 235 feet below sea level. In certain locations, the yearly rainfall varies from over 10 inches to roughly four inches. Summer temperatures are high, averaging 90°F (32°C) and rising as high as 125°F (52°C). Winter evenings are cool and days are moderate. The Atacama Desert, the driest of all the west coast deserts, is over 140,000 square miles in size and runs roughly 600 miles parallel to the Chilean coast. When oceanic air comes into contact with the Peruvian atmosphere, part of its moisture is lost. Its temperature is lowered by the current, and it loses the remaining heat as it passes over the coastal highlands. The majority of the Atacama is located in a valley behind the mountains. The average precipitation along the shore is roughly 0.4 inches, however, it comes in the form of fog rather than rain. Over the course of 21 years, Iquique in the north experienced an average of 0.06 inches of rain annually, with four years with no rain at all. For 19 years, Arica received less than 0.03 inches of rainfall annually(Nicholson, 1996).

8.8 AFRICAN DESERTS

There are two west coast deserts in Africa: The Atlantic Desert on the western edge of the Sahara and the Benguela Desert in southern Africa. Both are connected to the cold Canary Current. This is the Namib Desert and is roughly 19,300 square miles in size and is divided from the Kalahari by hills, with the exception of the south, where the two deserts meet. Rainfall averages 0.8 inches (20 mm) per year in Walvis Bay, which is located on the coast approximately halfway along the 932-mile length of the Namib and about 50 miles north of the tropic of Capricorn. Over the whole desert, the average annual precipitation is roughly two inches. During the night, low clouds frequently move in from the sea, bringing with them fog that quickly clears in the morning. The Namib is dry but not a particularly warm desert. Walvis Bay experiences 66–75°F (19–24°C) average temperatures. The Namib is mostly covered in gravel in the north, while to the south is a sand sea with the highest dunes in the world. An increasing amount of water vapor condenses as the atmospheric temperature drops. The air eventually gets so cold that practically no water vapor remains in it. The coldest places on Earth are Antarctica and Greenland. They are also among the driest due to their extreme cold. They are deserts in the Arctic. German scientists studied weather conditions while stationed on the Greenland ice cap during the winter of 1930–1931. They experienced the lowest average temperature of –52.9°F (–47.2°C) in February. The highest temperature recorded in July was 12.8°F (–3°C). Since then, numerous scientific missions have confirmed these temperatures on the ice cap. Ice that covers central Greenland is typically 5,000 feet thick, with a maximum thickness of over 8,000 feet. Although there seems to be an abundance of water, it is actually frozen water. The weather is unusually dry for a desert. Because of the air spaces between the ice crystals, snow occupies more space than liquid water; after melting, this quantity of snow is equal to roughly 0.3 inches of rain. The snow never melts since the temperature never goes above freezing. The ice sheet gradually thickens as a result of compression. It is currently thickening by roughly 203 mm/year. The ice sheet, sometimes known as the Greenland desert, is 708,069 square kilometers in size.

8.9 HUMUS

Humus, or the degraded remnants of plant and animal matter, is a mixture of mineral particles found in soil. Typically, chemical reactions that dissolve specific compounds found in the rocks produce the minerals, weakening the rock and enabling fragments to separate from it. Some of the compounds released in this manner are absorbed by plant roots; however, as the compounds are soluble in water, they are gradually washed to depths as the soil ages. This kind of drainage allows nutrients to build up below the surface, resulting in separate layers, or horizons. Bare rock is the starting point for soil formation, which matures into deep, layered soils that support plant life and the animals that coexist with it. The composition of the underlying rock and the kind of flora that develops there determine the properties of the developing soil. Farming existed throughout most of the northern Sahara as late as the Roman era(Gatto & Zerboni, 2015). These kinds of locations retain the ancient soils that were created under varying climate conditions. These soils could be over 75,000 years old, having formed before the Ice Age. There are fewer plants and very little organic matter in the soil when there is not enough water. The soil is infertile given the barren state of the land. Among all soil types, desert soils are some of the most fertile. The reason for this is that there isn't enough rain to remove the soluble nutrients. Very old soils, such those found in tropical rainforests, have practically lost their plant nutrients. Farmers who clear the forest to cultivate crops discover that they need to add a lot of fertilizer after a few years. Moisture is all that desert soil requires. If water is available, the crops will grow there. Deserts are frequently very cold at night and very hot during the day. Rocks are constantly expanding and contracting due to this large temperature range, which breaks them into fragments that are carried by the wind until they have been pulverized into tiny grains by bombardments. The grains are washed down slopes by rain, where they are released onto the plains below where they soon dry out. Silt and clay have a tendency to get deposited in hollows and on the beds of lakes. Some silt particles are carried by the wind past the desert. Loess soil deposits are created by windblown silt.

Approximately 360–290 million years ago, during the Carboniferous period, the region that is currently the Arabian Peninsula and the Sahara was submerged under the ocean. The seabed was covered with sand. Sandstone was created as the sediment was broken over time and the sand grains were cemented together. About 70 million years ago, the sea receded and then came back, leaving behind more sand. Sandstone made up a large portion of the land surface when it eventually rose from the sea, and it started to disintegrate. In the Sahara, a sand sea is referred to as an erg, and there are many. The Grand Eastern Erg in Algeria is one of the largest, covering almost 74,000 square miles. The ergs are located in areas of sand accumulation that resemble ocean basins. A gravel-covered stretch of elevated terrain spanning roughly 60 miles divides the Eastern and Western Ergs. Sand grains will flow down the sides of a dune if dry sand is poured upon its center. The sides of the dune won't get any steeper; instead, they will only get wider and higher. Never will they slope at a steeper degree than 35°. For dry sand, this is the angle of repose. The amount of sand available to form dunes and the direction of the wind both affect their appearance. The particles move beneath and around the stones by filling tiny air gaps below the surface. Occasionally, the wind adds extra dust to the surface, which is then washed away by the rain. The stones rise to the surface as the sand and dust are swept away beneath them very slowly. White sage, commonly known as winterfat (*Krascheninnikovia lanata*), is a member of the family Chenopodiaceae that grows in North America. It is a small bush with a thick layer of light-colored hairs covering it. While they are in the same family, *Agriophyllum gobicum* is an annual herb and its seeds are a main diet in Mongolia. The shrub known as *Haloxylon ammodendron* thrives in the sandy desert, reaching a height of roughly 10 feet. Saxaul plants grow in many of the less arid areas where they create forests that hold the soil together and stop erosion. Additionally, there are pea shrubs (*Caragana bungei* and *C. leucocephala*), which are small shrubs that dominate.

Figure 8.4 The saxaul

Source: CC BY-SA 2.5, https://commons.wikimedia.org/w/index.php?curid=295507,*Creative Commons License.*

REFERENCES

1. Aakala, T., & Keto-Tokoi, P. (2011). The old Norway spruce forests of northern boreal Fennoscandia are alive and well: A review of Sirén (1955). *Scandinavian Journal of Forest Research*, *26*(S10), 25–33. https://doi.org/10.1080/02827581.2011.517942
2. Aguadé, D., Poyatos, R., Rosas, T., & Martínez-Vilalta, J. (2015). Comparative Drought Responses of Quercus ilex L. and Pinus sylvestris L. in a Montane Forest Undergoing a Vegetation Shift. *Forests*, *6*(8), Article 8. https://doi.org/10.3390/f6082505
3. Aldenderfer, M. (2006). Modelling plateau peoples: The early human use of the world's high plateaux. *World Archaeology*, *38*(3), 357–370. https://doi.org/10.1080/00438240600813285
4. Alsherif, E. A., & Fadl, M. A. (2016). Floristic study of the Al-Shafa Highlands in Taif, Western Saudi Arabia. *Flora - Morphology, Distribution, Functional Ecology of Plants*, *225*, 20–29. https://doi.org/10.1016/j.flora.2016.09.004
5. Ameixa, O. M. C. C., & Sousa, A. I. (2020). Saltmarshes: Ecology, Opportunities, and Challenges. In W. Leal Filho, A. M. Azul, L. Brandli, A. Lange Salvia, & T. Wall (Eds.), *Life Below Water* (pp. 1–15). Springer International Publishing. https://doi.org/10.1007/978-3-319-71064-8_79-1
6. Arendt, C. A., Heikoop, J. M., Newman, B. D., Wilson, C. J., Wainwright, H., Kumar, J., Andersen, C. G., Wales, N. A., Dafflon, B., Cherry, J., & Wullschleger, S. D. (2022). Increased Arctic NO3− Availability as a Hydrogeomorphic Consequence of Permafrost Degradation and

Landscape Drying. *Nitrogen*, *3*(2), Article 2. https://doi.org/10.3390/nitrogen3020021

7. Arthur C. Trembanis, O. H. P., Hugo R. Valverde. (1999). Comparison of Beach Nourishment along the U.S. Atlantic, Great Lakes, Gulf of Mexico, and New England Shorelines. *Coastal Management*, *27*(4), 329–340. https://doi.org/10.1080/089207599263730
8. Arzhannikov, S. G., Ivanov, A. V., Arzhannikova, A. V., Demonterova, E. I., Jansen, J. D., Preusser, F., Kamenetsky, V. S., & Kamenetsky, M. B. (2018). Catastrophic events in the Quaternary outflow history of Lake Baikal. *Earth-Science Reviews*, *177*, 76–113. https://doi.org/10.1016/j.earscirev.2017.11.011
9. Aselmann, I., & Crutzen, P. J. (1989). Global distribution of natural freshwater wetlands and rice paddies, their net primary productivity, seasonality and possible methane emissions. *Journal of Atmospheric Chemistry*, *8*(4), 307–358. https://doi.org/10.1007/BF00052709
10. Asmus, H., Kneer, D., Pogoreutz, C., Blankenhorn, S., Jompa, J., Nurdin, N., & Priosambodo, D. (2022). 6—Ecology of seagrass beds in Sulawesi—Multifunctional key habitats at the risk of destruction. In T. C. Jennerjahn, T. Rixen, H. E. Irianto, & J. Samiaji (Eds.), *Science for the Protection of Indonesian Coastal Ecosystems (SPICE)* (pp. 201–250). Elsevier. https://doi.org/10.1016/B978-0-12-815050-4.00014-6
11. Ayarza, M., Rao, I., Vilela, L., Lascano, C., & Vera-Infanzón, R. (2022). Chapter Three - Soil carbon accumulation in crop-livestock systems in acid soil savannas of South America: A review. In D. L. Sparks (Ed.), *Advances in Agronomy* (Vol. 173, pp. 163–226). Academic Press. https://doi.org/10.1016/bs.agron.2022.02.003
12. Beamish, A., Raynolds, M. K., Epstein, H., Frost, G. V., Macander, M. J., Bergstedt, H., Bartsch, A., Kruse, S., Miles, V., Tanis, C. M., Heim, B., Fuchs, M., Chabrillat, S., Shevtsova, I., Verdonen, M., & Wagner, J. (2020). Recent trends and remaining challenges for optical remote sensing of Arctic tundra vegetation: A review and outlook. *Remote Sensing of Environment*, *246*, 111872. https://doi.org/10.1016/j.rse.2020.111872
13. Bergen, K. M., Hitztaler, S. K., Kharuk, V. I., Krankina, O. N., Loboda, T. V., Zhao, T., Shugart, H. H., & Sun, G. (2013). Human Dimensions of Environmental Change in Siberia. In P. Ya. Groisman & G. Gutman (Eds.), *Regional Environmental Changes in Siberia and Their Global Consequences* (pp. 251–302). Springer Netherlands. https://doi.org/10.1007/978-94-007-4569-8_7
14. Berón, M. A. (2016). Dunes, hills, waterholes, and saltpeter beds:

Attractors for human populations in western Pampa, Argentina. *Quaternary International*, *422*, 163–173. https://doi.org/10.1016/j.quaint.2016.03.001

15. Bertness, M. D. (1988). Peat Accumulation and the Success of Marsh Plants. *Ecology*, *69*(3), 703–713. https://doi.org/10.2307/1941018
16. Billings, W. D., & Mooney, H. A. (1968). The Ecology of Arctic and Alpine Plants. *Biological Reviews*, *43*(4), 481–529. https://doi.org/10.1111/j.1469-185X.1968.tb00968.x
17. Birks, H. J. B., & Willis, K. J. (2008). Alpines, trees, and refugia in Europe. *Plant Ecology & Diversity*, *1*(2), 147–160. https://doi.org/10.1080/17550870802349146
18. Bond, W. J., & Scott, A. C. (2010). Fire and the spread of flowering plants in the Cretaceous. *New Phytologist*, *188*(4), 1137–1150. https://doi.org/10.1111/j.1469-8137.2010.03418.x
19. Brotherton, S., Joyce, C. B., & Scharlemann, J. P. W. (2020). Global offtake of wild animals from wetlands: Critical issues for fish and birds. *Hydrobiologia*, *847*(7), 1631–1649. https://doi.org/10.1007/s10750-020-04188-z
20. Bussmann, R. W. (2002). Vegetation Ecology and Regeneration of Tropical Mountain Forests. In R. S. Ambasht & N. K. Ambasht (Eds.), *Modern Trends in Applied Terrestrial Ecology* (pp. 195–223). Springer US. https://doi.org/10.1007/978-1-4615-0223-4_11
21. Cabido, M., Pons, E., Cantero, J. J., Lewis, J. P., & Anton, A. (2008). Photosynthetic pathway variation among C4 grasses along a precipitation gradient in Argentina. *Journal of Biogeography*, *35*(1), 131–140. https://doi.org/10.1111/j.1365-2699.2007.01760.x
22. Campbell, D. I., & Williamson, J. L. (1997). Evaporation from a raised peat bog. *Journal of Hydrology*, *193*(1), 142–160. https://doi.org/10.1016/S0022-1694(96)03149-6
23. Capellán-Pérez, I., de Castro, C., & Arto, I. (2017). Assessing vulnerabilities and limits in the transition to renewable energies: Land requirements under 100% solar energy scenarios. *Renewable and Sustainable Energy Reviews*, *77*, 760–782. https://doi.org/10.1016/j.rser.2017.03.137
24. Catenazzi, A., & A. Donnelly, M. (2007). The Ulva connection: Marine algae subsidize terrestrial predators in coastal Peru. *Oikos*, *116*(1), 75–86. https://doi.org/10.1111/j.2006.0030-1299.15230.x
25. Chakraborty, S. K., Sanyal, P., & Ray, R. (2023). Diversity and Classification of Wetlands in International and National Perspectives. In S. K. Chakraborty, P. Sanyal, & R. Ray (Eds.), *Wetlands Ecology: Eco-*

biological uniqueness of a Ramsar site (East Kolkata Wetlands, India) (pp. 167–226). Springer International Publishing. https://doi.org/10.1007/978-3-031-09253-4_3

26. Chapin III, F. S., Shaver, G. R., Giblin, A. E., Nadelhoffer, K. J., & Laundre, J. A. (1995). Responses of Arctic Tundra to Experimental and Observed Changes in Climate. *Ecology*, *76*(3), 694–711. https://doi.org/10.2307/1939337
27. Chernov, N. M., Yuri. (2000). Biodiversity of Terrestrial Ecosystems. In *The Arctic*. Routledge.
28. Clark, D. H., & Gillespie, A. R. (1997). Timing and significance of Late-glacial and Holocene cirque glaciation in the Sierra Nevada, California. *Quaternary International*, *38–39*, 21–38. https://doi.org/10.1016/S1040-6182(96)00024-9
29. Clarke, J., Brooks, N., Banning, E. B., Bar-Matthews, M., Campbell, S., Clare, L., Cremaschi, M., di Lernia, S., Drake, N., Gallinaro, M., Manning, S., Nicoll, K., Philip, G., Rosen, S., Schoop, U.-D., Tafuri, M. A., Weninger, B., & Zerboni, A. (2016). Climatic changes and social transformations in the Near East and North Africa during the 'long' 4th millennium BC: A comparative study of environmental and archaeological evidence. *Quaternary Science Reviews*, *136*, 96–121. https://doi.org/10.1016/j.quascirev.2015.10.003
30. Colyn, R. B., Campbell, A. M., & Smit-Robinson, H. A. (2017). The application of camera trapping to assess Rallidae species richness within palustrine wetland habitat in South Africa. *Ostrich*, *88*(3), 235–245. https://doi.org/10.2989/00306525.2017.1292562
31. Cordell, J. R., Lawrence, D. J., Ferm, N. C., Tear, L. M., Smith, S. S., & Herwig, R. P. (2009). Factors influencing densities of non-indigenous species in the ballast water of ships arriving at ports in Puget Sound, Washington, United States. *Aquatic Conservation: Marine and Freshwater Ecosystems*, *19*(3), 322–343. https://doi.org/10.1002/aqc.986
32. Culp, J. M., Luiker, E. A., Brua, R. B., Musetta-Lambert, J. L., Halliwell, D. B., & Lento, J. (2023). Chapter 18—Mackenzie River Basin. In M. D. Delong, T. D. Jardine, A. C. Benke, & C. E. Cushing (Eds.), *Rivers of North America (Second Edition)* (pp. 746–792). Academic Press. https://doi.org/10.1016/B978-0-12-818847-7.00017-3
33. Ćurić, M., & Spiridonov, V. (2023). Atmospheric Motion. In M. Ćurić & V. Spiridonov (Eds.), *History of Meteorology* (pp. 251–265). Springer Nature Switzerland. https://doi.org/10.1007/978-3-031-45032-7_12
34. Dembitsky, V. M., Poovarodom, S., Leontowicz, H., Leontowicz, M.,

Vearasilp, S., Trakhtenberg, S., & Gorinstein, S. (2011). The multiple nutrition properties of some exotic fruits: Biological activity and active metabolites. *Food Research International*, *44*(7), 1671–1701. https://doi.org/10.1016/j.foodres.2011.03.003

35. Den Hartog, C., & Van Der Velde, G. (1988). Structural Aspects of Aquatic Plant Communities. In J. J. Symoens (Ed.), *Vegetation of inland waters* (pp. 113–153). Springer Netherlands. https://doi.org/10.1007/978-94-009-3087-2_4
36. Dirmeyer, P. A., & Shukla, J. (1996). The effect on regional and global climate of expansion of the world's deserts. *Quarterly Journal of the Royal Meteorological Society*, *122*(530), 451–482. https://doi.org/10.1002/qj.49712253008
37. Drabkova, V. G., Rutnyantsev, V. A., Sergeeva, L. V., & Slepukhina, T. D. (1996). Ecological problems of Lake Ladoga: Causes and solutions. *Hydrobiologia*, *322*(1), 1–7. https://doi.org/10.1007/BF00031798
38. Dredge, L. A., & Dyke, L. D. (2020). Landscapes and Landforms of the Hudson Bay Lowlands. In O. Slaymaker & N. Catto (Eds.), *Landscapes and Landforms of Eastern Canada* (pp. 211–227). Springer International Publishing. https://doi.org/10.1007/978-3-030-35137-3_8
39. Eidt, R. C. (1969). The Climatology of South America. In E. J. Fittkau, J. Illies, H. Klinge, G. H. Schwabe, & H. Sioli (Eds.), *Biogeography and Ecology in South America* (pp. 54–81). Springer Netherlands. https://doi.org/10.1007/978-94-011-9731-1_3
40. Emeis, K.-C., van Beusekom, J., Callies, U., Ebinghaus, R., Kannen, A., Kraus, G., Kröncke, I., Lenhart, H., Lorkowski, I., Matthias, V., Möllmann, C., Pätsch, J., Scharfe, M., Thomas, H., Weisse, R., & Zorita, E. (2015). The North Sea—A shelf sea in the Anthropocene. *Journal of Marine Systems*, *141*, 18–33. https://doi.org/10.1016/j.jmarsys.2014.03.012
41. Emmerton, C. A., Lesack, L. F. W., & Marsh, P. (2007). Lake abundance, potential water storage, and habitat distribution in the Mackenzie River Delta, western Canadian Arctic. *Water Resources Research*, *43*(5). https://doi.org/10.1029/2006WR005139
42. Epstein, H. E., Beringer, J., Gould, W. A., Lloyd, A. H., Thompson, C. D., Chapin III, F. S., Michaelson, G. J., Ping, C. L., Rupp, T. S., & Walker, D. A. (2004). The nature of spatial transitions in the Arctic. *Journal of Biogeography*, *31*(12), 1917–1933. https://doi.org/10.1111/j.1365-2699.2004.01140.x
43. Eugster, W., Rouse, W. R., Pielke Sr, R. A., Mcfadden, J. P., Baldocchi, D. D., Kittel, T. G. F., Chapin III, F. S., Liston, G. E., Vidale, P. L.,

Vaganov, E., & Chambers, S. (2000). Land–atmosphere energy exchange in Arctic tundra and boreal forest: Available data and feedbacks to climate. *Global Change Biology*, *6*(S1), 84–115. https://doi.org/10.1046/j.1365-2486.2000.06015.x

44. Ezcurra, E., Martínez-Berdeja, A., & Villanueva-Almanza, L. (2020). The Evolution of North American Deserts and the Uniqueness of Cuatro Ciénegas. In M. C. Mandujano, I. Pisanty, & L. E. Eguiarte (Eds.), *Plant Diversity and Ecology in the Chihuahuan Desert: Emphasis on the Cuatro Ciénegas Basin* (pp. 45–60). Springer International Publishing. https://doi.org/10.1007/978-3-030-44963-6_3
45. Fedorov, V. B., Goropashnaya, A. V., Boeskorov, G. G., & Cook, J. A. (2008). Comparative phylogeography and demographic history of the wood lemming (Myopus schisticolor): Implications for late Quaternary history of the taiga species in Eurasia. *Molecular Ecology*, *17*(2), 598–610. https://doi.org/10.1111/j.1365-294X.2007.03595.x
46. Fernández-Delgado, C. (1989). Life-history patterns of the salt-marsh killifish *Fundulus heteroclitus* (L.) introduced in the estuary of the guadalquivir river (South West Spain). *Estuarine, Coastal and Shelf Science*, *29*(6), 573–582. https://doi.org/10.1016/0272-7714(89)90011-5
47. Foster, P. (2001). The potential negative impacts of global climate change on tropical montane cloud forests. *Earth-Science Reviews*, *55*(1), 73–106. https://doi.org/10.1016/S0012-8252(01)00056-3
48. Friedland, K. D., Langan, J. A., Large, S. I., Selden, R. L., Link, J. S., Watson, R. A., & Collie, J. S. (2020). Changes in higher trophic level productivity, diversity and niche space in a rapidly warming continental shelf ecosystem. *Science of The Total Environment*, *704*, 135270. https://doi.org/10.1016/j.scitotenv.2019.135270
49. Garrett, B. C., Schenter, G. K., & Morita, A. (2006). Molecular Simulations of the Transport of Molecules across the Liquid/Vapor Interface of Water. *Chemical Reviews*, *106*(4), 1355–1374. https://doi.org/10.1021/cr040370w
50. Gatto, M. C., & Zerboni, A. (2015). Holocene Supra-Regional Environmental Changes as Trigger for Major Socio-Cultural Processes in Northeastern Africa and the Sahara. *African Archaeological Review*, *32*(2), 301–333. https://doi.org/10.1007/s10437-015-9191-x
51. George, J. D. (2012). Reef-Associated Macroinvertebrates of the SE Gulf. In B. M. Riegl & S. J. Purkis (Eds.), *Coral Reefs of the Gulf: Adaptation to Climatic Extremes* (pp. 253–308). Springer Netherlands. https://doi.org/10.1007/978-94-007-3008-3_13

52. Geresdi, I., Sarkadi, N., & Thompson, G. (2014). Effect of the accretion by water drops on the melting of snowflakes. *Atmospheric Research*, *149*, 96–110. https://doi.org/10.1016/j.atmosres.2014.06.001
53. Goudie, A. S., Goudie, A. M., & Viles, H. A. (2021). The distribution and nature of star dunes: A global analysis. *Aeolian Research*, *50*, 100685. https://doi.org/10.1016/j.aeolia.2021.100685
54. Gough, W. A., & He, D. (2015). Diurnal temperature asymmetries and fog at Churchill, Manitoba. *Theoretical and Applied Climatology*, *121*(1), 113–119. https://doi.org/10.1007/s00704-014-1227-7
55. Grabherr, G., Gottfried, M., & Pauli, H. (2010). Climate Change Impacts in Alpine Environments. *Geography Compass*, *4*(8), 1133–1153. https://doi.org/10.1111/j.1749-8198.2010.00356.x
56. Green, T. G. A., Sancho, L. G., & Pintado, A. (2011). Ecophysiology of Desiccation/Rehydration Cycles in Mosses and Lichens. In U. Lüttge, E. Beck, & D. Bartels (Eds.), *Plant Desiccation Tolerance* (pp. 89–120). Springer. https://doi.org/10.1007/978-3-642-19106-0_6
57. Griffin, P. J., & Ali, S. H. (2014). Managing transboundary wetlands: The Ramsar Convention as a means of ecological diplomacy. *Journal of Environmental Studies and Sciences*, *4*(3), 230–239. https://doi.org/10.1007/s13412-014-0173-0
58. Groisman, P. Ya., Blyakharchuk, T. A., Chernokulsky, A. V., Arzhanov, M. M., Marchesini, L. B., Bogdanova, E. G., Borzenkova, I. I., Bulygina, O. N., Karpenko, A. A., Karpenko, L. V., Knight, R. W., Khon, V. Ch., Korovin, G. N., Meshcherskaya, A. V., Mokhov, I. I., Parfenova, E. I., Razuvaev, V. N., Speranskaya, N. A., Tchebakova, N. M., & Vygodskaya, N. N. (2013). Climate Changes in Siberia. In P. Ya. Groisman & G. Gutman (Eds.), *Regional Environmental Changes in Siberia and Their Global Consequences* (pp. 57–109). Springer Netherlands. https://doi.org/10.1007/978-94-007-4569-8_3
59. Guo, W., Rees, G., & Hofgaard, A. (2020). Delineation of the forest-tundra ecotone using texture-based classification of satellite imagery. *International Journal of Remote Sensing*, *41*(16), 6384–6408. https://doi.org/10.1080/01431161.2020.1734254
60. Hamdan, M. A., Refaat, A. A., & Abdel Wahed, M. (2016). Morphologic characteristics and migration rate assessment of barchan dunes in the Southeastern Western Desert of Egypt. *Geomorphology*, *257*, 57–74. https://doi.org/10.1016/j.geomorph.2015.12.026
61. Hansson, J., Berndes, G., Englund, O., De Freitas, F. L. M., & Sparovek, G. (2019). How is biodiversity protection influencing the potential for

bioenergy feedstock production on grasslands? *GCB Bioenergy*, *11*(3), 517–538. https://doi.org/10.1111/gcbb.12568

62. Harborne, A. R., Mumby, P. J., Micheli, F., Perry, C. T., Dahlgren, C. P., Holmes, K. E., & Brumbaugh, D. R. (2006). The Functional Value of Caribbean Coral Reef, Seagrass and Mangrove Habitats to Ecosystem Processes. In *Advances in Marine Biology* (Vol. 50, pp. 57–189). Academic Press. https://doi.org/10.1016/S0065-2881(05)50002-6
63. Hereford, R., Webb, R. H., & Longpré, C. I. (2006). Precipitation history and ecosystem response to multidecadal precipitation variability in the Mojave Desert region, 1893–2001. *Journal of Arid Environments*, *67*, 13–34. https://doi.org/10.1016/j.jaridenv.2006.09.019
64. Hoekstra, B. R., Neill, C., & Kennedy, C. D. (2020). Trends in the Massachusetts cranberry industry create opportunities for the restoration of cultivated riparian wetlands. *Restoration Ecology*, *28*(1), 185–195. https://doi.org/10.1111/rec.13037
65. Hu, S., Niu, Z., Chen, Y., Li, L., & Zhang, H. (2017). Global wetlands: Potential distribution, wetland loss, and status. *Science of The Total Environment*, *586*, 319–327. https://doi.org/10.1016/j.scitotenv.2017.02.001
66. Huntley, B. J. (2023). Adaptations to Life in the Namib Desert. In B. J. Huntley (Ed.), *Ecology of Angola: Terrestrial Biomes and Ecoregions* (pp. 249–274). Springer International Publishing. https://doi.org/10.1007/978-3-031-18923-4_11
67. Infante Mata, D., Moreno-Casasola, P., Madero-Vega, C., Castillo-Campos, G., & Warner, B. G. (2011). Floristic composition and soil characteristics of tropical freshwater forested wetlands of Veracruz on the coastal plain of the Gulf of Mexico. *Forest Ecology and Management*, *262*(8), 1514–1531. https://doi.org/10.1016/j.foreco.2011.06.053
68. Janzen, D. H., & Schoener, T. W. (1968). Differences in Insect Abundance and Diversity Between Wetter and Drier Sites During a Tropical Dry Season. *Ecology*, *49*(1), 96–110. https://doi.org/10.2307/1933565
69. Jarvis, A., & Mulligan, M. (2011). The climate of cloud forests. *Hydrological Processes*, *25*(3), 327–343. https://doi.org/10.1002/hyp.7847
70. Jennings, S. J. A., & Hambrey, M. J. (2021). Structures and Deformation in Glaciers and Ice Sheets. *Reviews of Geophysics*, *59*(3), e2021RG000743. https://doi.org/10.1029/2021RG000743
71. Jewell, J., Vinichenko, V., Nacke, L., & Cherp, A. (2019). Prospects for powering past coal. *Nature Climate Change*, *9*(8), Article 8. https://doi.org/10.1038/s41558-019-0509-6

72. Jobst, A. M., Kingston, D. G., Cullen, N. J., & Sirguey, P. (2017). Combining thin-plate spline interpolation with a lapse rate model to produce daily air temperature estimates in a data-sparse alpine catchment. *International Journal of Climatology*, *37*(1), 214–229. https://doi.org/10.1002/joc.4699

73. Juda, L. (1993). Ocean policy, multi-use management, and the cumulative impact of piecemeal change: The case of the United States outer continental shelf. *Ocean Development & International Law*, *24*(4), 355–376. https://doi.org/10.1080/00908329309546017

74. Junk, W. J., An, S., Finlayson, C. M., Gopal, B., Květ, J., Mitchell, S. A., Mitsch, W. J., & Robarts, R. D. (2013). Current state of knowledge regarding the world's wetlands and their future under global climate change: A synthesis. *Aquatic Sciences*, *75*(1), 151–167. https://doi.org/10.1007/s00027-012-0278-z

75. Junk, W. J., Piedade, M. T. F., Wittmann, F., & Schöngart, J. (2011). The Role of Floodplain Forests in an Integrated Sustainable Management Concept of the Natural Resources of the Central Amazonian Várzea. In W. J. Junk, M. T. F. Piedade, F. Wittmann, J. Schöngart, & P. Parolin (Eds.), *Amazonian Floodplain Forests: Ecophysiology, Biodiversity and Sustainable Management* (pp. 485–509). Springer Netherlands. https://doi.org/10.1007/978-90-481-8725-6_23

76. Kauss, P. B. (1991). Biota of the St. Marys River: Habitat evaluation and environmental assessment. In M. Munawar & T. Edsall (Eds.), *Environmental Assessment and Habitat Evaluation of the Upper Great Lakes Connecting Channels* (pp. 1–35). Springer Netherlands. https://doi.org/10.1007/978-94-011-3144-5_1

77. Kharuk, V. I., Petrov, I. A., Golyukov, A. S., Dvinskaya, M. L., Im, S. T., & Shushpanov, A. S. (2023). Larch growth across thermal and moisture gradients in the Siberian Mountains. *Journal of Mountain Science*, *20*(1), 101–114. https://doi.org/10.1007/s11629-022-7433-3

78. Konrad, W., Roth-Nebelsick, A., & Traiser, C. (2023). High Productivity at High Latitudes? Photosynthesis and Leaf Ecophysiology in Arctic Forests of the Eocene. *Paleoceanography and Paleoclimatology*, *38*(8), e2023PA004685. https://doi.org/10.1029/2023PA004685

79. Körner, C. (2021). Alpine treelines. In C. Körner (Ed.), *Alpine Plant Life: Functional Plant Ecology of High Mountain Ecosystems* (pp. 141–173). Springer International Publishing. https://doi.org/10.1007/978-3-030-59538-8_7

80. Körner, C., & Paulsen, J. (2004). A world-wide study of high altitude treeline temperatures. *Journal of Biogeography*, *31*(5), 713–732. https://doi.org/10.1111/j.1365-2699.2003.01043.x

81. Kueppers, U., & Beier, C. (Eds.). (2018). *Volcanoes of the Azores: Revealing the Geological Secrets of the Central Northern Atlantic Islands*. Springer. https://doi.org/10.1007/978-3-642-32226-6
82. Kuhle, M. (1998). Reconstruction of the 2.4 million km2 late Pleistocene ice sheet on the Tibetan Plateau and its impact on the global climate. *Quaternary International*, *45–46*, 71–108. https://doi.org/10.1016/S1040-6182(97)00008-6
83. Laiho, R. (2006). Decomposition in peatlands: Reconciling seemingly contrasting results on the impacts of lowered water levels. *Soil Biology and Biochemistry*, *38*(8), 2011–2024. https://doi.org/10.1016/j.soilbio.2006.02.017
84. Lamprey, H. F. (1964). Estimation of the Large Mammal Densities, Biomass and Energy Exchange in the Tarangire Game Reserve and the Masai Steppe in Tanganyika. *African Journal of Ecology*, *2*(1), 1–46. https://doi.org/10.1111/j.1365-2028.1964.tb00194.x
85. Lan, H. X., Hu, R. L., Yue, Z. Q., Lee, C. F., & Wang, S. J. (2003). Engineering and geological characteristics of granite weathering profiles in South China. *Journal of Asian Earth Sciences*, *21*(4), 353–364. https://doi.org/10.1016/S1367-9120(02)00020-2
86. Lee, D., Lee, W., Jordan, G., & Barreda, V. (2016). The Cenozoic history of New Zealand temperate rainforests: Comparisons with southern Australia and South America. *New Zealand Journal of Botany*, *54*(2), 100–127. https://doi.org/10.1080/0028825X.2016.1144623
87. Lemly, A. D., & Dimmick, J. F. (1982). Structure and dynamics of zooplankton communities in the littoral zone of some North Carolina lakes. *Hydrobiologia*, *88*(3), 299–307. https://doi.org/10.1007/BF00008511
88. Leonardi, N., Carnacina, I., Donatelli, C., Ganju, N. K., Plater, A. J., Schuerch, M., & Temmerman, S. (2018). Dynamic interactions between coastal storms and salt marshes: A review. *Geomorphology*, *301*, 92–107. https://doi.org/10.1016/j.geomorph.2017.11.001
89. Letelier, R. M., Karl, D. M., Abbott, M. R., & Bidigare, R. R. (2004). Light driven seasonal patterns of chlorophyll and nitrate in the lower euphotic zone of the North Pacific Subtropical Gyre. *Limnology and Oceanography*, *49*(2), 508–519. https://doi.org/10.4319/lo.2004.49.2.0508
90. Liang, M., Sugimoto, A., Tei, S., Bragin, I. V., Takano, S., Morozumi, T., Shingubara, R., Maximov, T. C., Kiyashko, S. I., Velivetskaya, T. A., & Ignatiev, A. V. (2014). Importance of soil moisture and N availability to larch growth and distribution in the Arctic taiga-tundra boundary

ecosystem, northeastern Siberia. *Polar Science*, *8*(4), 327–341. https://doi.org/10.1016/j.polar.2014.07.008

91. Lisetskii, F. N., Smekalova, T. N., & Marinina, O. A. (2016). Biogeochemical features of fallow lands in the steppe zone. *Contemporary Problems of Ecology*, *9*(3), 366–375. https://doi.org/10.1134/S1995425516030094
92. Lovett, G. M., & Goodale, C. L. (2011). A New Conceptual Model of Nitrogen Saturation Based on Experimental Nitrogen Addition to an Oak Forest. *Ecosystems*, *14*(4), 615–631. https://doi.org/10.1007/s10021-011-9432-z
93. Lundell, A., Cousins, S. A. O., & Eriksson, O. (2015). Population size and reproduction in the declining endangered forest plant Chimaphila umbellata in Sweden. *Folia Geobotanica*, *50*(1), 13–23. https://doi.org/10.1007/s12224-015-9212-1
94. Manning, A. H., Verplanck, P. L., Caine, J. S., & Todd, A. S. (2013). Links between climate change, water-table depth, and water chemistry in a mineralized mountain watershed. *Applied Geochemistry*, *37*, 64–78. https://doi.org/10.1016/j.apgeochem.2013.07.002
95. Marengo, J. A., Tomasella, J., Soares, W. R., Alves, L. M., & Nobre, C. A. (2012). Extreme climatic events in the Amazon basin. *Theoretical and Applied Climatology*, *107*(1), 73–85. https://doi.org/10.1007/s00704-011-0465-1
96. Markle, D. F., Reithel, S. A., Crandall, J., Wood, T., Tyler, T. J., Terwilliger, M., & Simon, D. C. (2009). Larval Fish Transport and Retention and the Importance of Location for Juvenile Fish Recruitment in Upper Klamath Lake, Oregon. *Transactions of the American Fisheries Society*, *138*(2), 328–347. https://doi.org/10.1577/T07-274.1
97. Martínez, M. L., Intralawan, A., Vázquez, G., Pérez-Maqueo, O., Sutton, P., & Landgrave, R. (2007). The coasts of our world: Ecological, economic and social importance. *Ecological Economics*, *63*(2), 254–272. https://doi.org/10.1016/j.ecolecon.2006.10.022
98. Mayden, R. L., & Chen, W.-J. (2010). The world's smallest vertebrate species of the genus *Paedocypris*: A new family of freshwater fishes and the sister group to the world's most diverse clade of freshwater fishes (Teleostei: Cypriniformes). *Molecular Phylogenetics and Evolution*, *57*(1), 152–175. https://doi.org/10.1016/j.ympev.2010.04.008
99. Mayorga, E., Logsdon, M. G., Ballester, M. V. R., & Richey, J. E. (2005). Estimating cell-to-cell land surface drainage paths from digital channel networks, with an application to the Amazon basin. *Journal of Hydrology*, *315*(1), 167–182. https://doi.org/10.1016/j.jhydrol.2005.03.023

100. McGlue, M. M., & Woolery, E. W. (2022). High-resolution seismic profiling reveals basin floor morphology, sedimentary processes, and shallow stratigraphy at Convict Lake (California, USA). *Quaternary International*, *621*, 37–49. https://doi.org/10.1016/j.quaint.2020.09.014

101. McKee, K. L., & Patrick, W. H. (1988). The relationship of smooth cordgrass (Spartina alterniflora) to tidal datums: A review. *Estuaries*, *11*(3), 143–151. https://doi.org/10.2307/1351966

102. Meixner, M. J., Lüter, C., Eckert, C., Itskovich, V., Janussen, D., von Rintelen, T., Bohne, A. V., Meixner, J. M., & Hess, W. R. (2007). Phylogenetic analysis of freshwater sponges provide evidence for endemism and radiation in ancient lakes. *Molecular Phylogenetics and Evolution*, *45*(3), 875–886. https://doi.org/10.1016/j.ympev.2007.09.007

103. Miller, D., Poucher, S., & Coiro, L. (2002). Determination of lethal dissolved oxygen levels for selected marine and estuarine fishes, crustaceans, and a bivalve. *Marine Biology*, *140*(2), 287–296. https://doi.org/10.1007/s002270100702

104. Mitchell, D., Henschel, J. R., Hetem, R. S., Wassenaar, T. D., Strauss, W. M., Hanrahan, S. A., & Seely, M. K. (2020). Fog and fauna of the Namib Desert: Past and future. *Ecosphere*, *11*(1), e02996. https://doi.org/10.1002/ecs2.2996

105. Moorhead, K. K., & Cook, A. E. (1992). A comparison of hydric soils, wetlands, and land Use in coastal North Carolina. *Wetlands*, *12*(2), 99–105. https://doi.org/10.1007/BF03160591

106. Moschella, P. S., Abbiati, M., Åberg, P., Airoldi, L., Anderson, J. M., Bacchiocchi, F., Bulleri, F., Dinesen, G. E., Frost, M., Gacia, E., Granhag, L., Jonsson, P. R., Satta, M. P., Sundelöf, A., Thompson, R. C., & Hawkins, S. J. (2005). Low-crested coastal defence structures as artificial habitats for marine life: Using ecological criteria in design. *Coastal Engineering*, *52*(10), 1053–1071. https://doi.org/10.1016/j.coastaleng.2005.09.014

107. Murton, J. B. (2021). What and where are periglacial landscapes? *Permafrost and Periglacial Processes*, *32*(2), 186–212. https://doi.org/10.1002/ppp.2102

108. Nelson, C. H. (2021). *Witness To A Changing Earth: A Geologist's Journey Learning About Natural and Human-caused Global Change*. Springer International Publishing. https://doi.org/10.1007/978-3-030-71811-4

109. Nicholls, R. J., & Hoozemans, F. M. J. (1996). The Mediterranean: Vulnerability to coastal implications of climate change. *Ocean & Coastal Management*, *31*(2), 105–132. https://doi.org/10.1016/S0964-5691(96)00037-3

110. Nicholson, S. E. (1996). A Review of Climate Dynamics and Climate Variability in Eastern Africa. In *Limnology, Climatology and Paleoclimatology of the East African Lakes*. Routledge.
111. Niwa, M., Shimada, K., Aoki, K., & Ishimaru, T. (2016). Microscopic features of quartz and clay particles from fault gouges and infilled fractures in granite: Discriminating between active and inactive faulting. *Engineering Geology*, *210*, 180–196. https://doi.org/10.1016/j.enggeo.2016.06.013
112. Nyström, M., Norström, A. V., Blenckner, T., de la Torre-Castro, M., Eklöf, J. S., Folke, C., Österblom, H., Steneck, R. S., Thyresson, M., & Troell, M. (2012). Confronting Feedbacks of Degraded Marine Ecosystems. *Ecosystems*, *15*(5), 695–710. https://doi.org/10.1007/s10021-012-9530-6
113. Oren, A. (2008). Microbial life at high salt concentrations: Phylogenetic and metabolic diversity. *Saline Systems*, *4*(1), 2. https://doi.org/10.1186/1746-1448-4-2
114. Pages, P. I. W. G. of. (2016). Interglacials of the last 800,000 years. *Reviews of Geophysics*, *54*(1), 162–219. https://doi.org/10.1002/2015RG000482
115. Pauly, D., Graham, W., Libralato, S., Morissette, L., & Palomares, M. L. D. (2009). Jellyfish in ecosystems, online databases, and ecosystem models. In K. A. Pitt & J. E. Purcell (Eds.), *Jellyfish Blooms: Causes, Consequences, and Recent Advances: Proceedings of the Second International Jellyfish Blooms Symposium, held at the Gold Coast, Queensland, Australia, 24–27 June, 2007* (pp. 67–85). Springer Netherlands. https://doi.org/10.1007/978-1-4020-9749-2_5
116. Pauly, D., & Liang, C. (2020). The fisheries of the South China Sea: Major trends since 1950. *Marine Policy*, *121*, 103584. https://doi.org/10.1016/j.marpol.2019.103584
117. Pennington, B. T., Sturt, F., Wilson, P., Rowland, J., & Brown, A. G. (2017). The fluvial evolution of the Holocene Nile Delta. *Quaternary Science Reviews*, *170*, 212–231. https://doi.org/10.1016/j.quascirev.2017.06.017
118. Pfadenhauer, J. S., & Klötzli, F. A. (2020). Zonal Vegetation of the Tropical Zone with Summer Rain. In J. S. Pfadenhauer & F. A. Klötzli (Eds.), *Global Vegetation: Fundamentals, Ecology and Distribution* (pp. 179–258). Springer International Publishing. https://doi.org/10.1007/978-3-030-49860-3_3
119. Pielke, R. A., & Vidale, P. L. (1995). The boreal forest and the polar front. *Journal of Geophysical Research: Atmospheres*, *100*(D12), 25755–25758. https://doi.org/10.1029/95JD02418

120. Pimentel, D., Bailey, O., Kim, P., Mullaney, E., Calabrese, J., Walman, L., Nelson, F., & Yao, X. (1999). Will Limits of the Earth's Resources Control Human Numbers? *Environment, Development and Sustainability*, *1*(1), 19–39. https://doi.org/10.1023/A:1010008112119

121. Pócs, T. (1982). Tropical Forest Bryophytes. In A. J. E. Smith (Ed.), *Bryophyte Ecology* (pp. 59–104). Springer Netherlands. https://doi.org/10.1007/978-94-009-5891-3_3

122. Pollard, W. (2018). Chapter 15—Periglacial Processes in Glacial Environments. In J. Menzies & J. J. M. van der Meer (Eds.), *Past Glacial Environments (Second Edition)* (pp. 537–564). Elsevier. https://doi.org/10.1016/B978-0-08-100524-8.00016-6

123. Pratolongo, P., Leonardi, N., Kirby, J. R., & Plater, A. (2019). Chapter 3 - Temperate Coastal Wetlands: Morphology, Sediment Processes, and Plant Communities. In G. M. E. Perillo, E. Wolanski, D. R. Cahoon, & C. S. Hopkinson (Eds.), *Coastal Wetlands (Second Edition)* (pp. 105–152). Elsevier. https://doi.org/10.1016/B978-0-444-63893-9.00003-4

124. Qin, Y., Xiao, X., Dong, J., Zhou, Y., Wang, J., Doughty, R. B., Chen, Y., Zou, Z., & Moore, B. (2017). Annual dynamics of forest areas in South America during 2007–2010 at 50-m spatial resolution. *Remote Sensing of Environment*, *201*, 73–87. https://doi.org/10.1016/j.rse.2017.09.005

125. Rabalais, N. N., Smith, L. E., Harper Jr., D. E., & Justic, D. (2001). Effects of Seasonal Hypoxia on Continental Shelf Benthos. In *Coastal Hypoxia: Consequences for Living Resources and Ecosystems* (pp. 211–240). American Geophysical Union (AGU). https://doi.org/10.1029/CE058p0211

126. Rahel, F. J. (1984). Factors Structuring Fish Assemblages Along a Bog Lake Successional Gradient. *Ecology*, *65*(4), 1276–1289. https://doi.org/10.2307/1938333

127. Raihana, A. R. N., Marikkar, J. M. N., Amin, I., & Shuhaimi, M. (2015). A Review on Food Values of Selected Tropical Fruits' Seeds. *International Journal of Food Properties*, *18*(11), 2380–2392. https://doi.org/10.1080/10942912.2014.980946

128. Rajan, S. (2018). The Legal Continental Shelf: Geosciences at sea with UNCLOS. *Journal of the Geological Society of India*, *92*(2), 131–133. https://doi.org/10.1007/s12594-018-0970-2

129. Rana, R., Langenfeld-Heyser, R., Finkeldey, R., & Polle, A. (2009). Functional anatomy of five endangered tropical timber wood species of the family Dipterocarpaceae. *Trees*, *23*(3), 521–529. https://doi.org/10.1007/s00468-008-0298-4

130. Ranson, K. J., Montesano, P. M., & Nelson, R. (2011). Object-based mapping of the circumpolar taiga–tundra ecotone with MODIS tree cover. *Remote Sensing of Environment*, *115*(12), 3670–3680. https://doi.org/10.1016/j.rse.2011.09.006
131. Rapp, D. (2019). *Ice Ages and Interglacials: Measurements, Interpretation, and Models*. Springer International Publishing. https://doi.org/10.1007/978-3-030-10466-5
132. Reeves, J. S. (1930). The Codification of the Law of Territorial Waters. *American Journal of International Law*, *24*(3), 486–499. https://doi.org/10.2307/2189681
133. Reis, R. E., Albert, J. S., Di Dario, F., Mincarone, M. M., Petry, P., & Rocha, L. A. (2016). Fish biodiversity and conservation in South America. *Journal of Fish Biology*, *89*(1), 12–47. https://doi.org/10.1111/jfb.13016
134. Reynolds, C. S. (1994). The long, the short and the stalled: On the attributes of phytoplankton selected by physical mixing in lakes and rivers. *Hydrobiologia*, *289*(1), 9–21. https://doi.org/10.1007/BF00007405
135. Richardson, A. D., & Friedland, A. J. (2009). A Review of the Theories to Explain Arctic and Alpine Treelines Around the World. *Journal of Sustainable Forestry*, *28*(1–2), 218–242. https://doi.org/10.1080/10549810802626456
136. Rodov, S. B.-Y., Victor. (2002). Transpiration and Water Stress. In *Postharvest Physiology and Pathology of Vegetables* (2nd ed.). CRC Press.
137. Salmon, M., & Hyatt, G. W. (1983). Spatial and temporal aspects of reproduction in North Carolina fiddler crabs (*Uca pugilator* Bosc). *Journal of Experimental Marine Biology and Ecology*, *70*(1), 21–43. https://doi.org/10.1016/0022-0981(83)90146-6
138. Salzmann, U., Haywood, A. M., Lunt, D. J., Valdes, P. J., & Hill, D. J. (2008). A new global biome reconstruction and data-model comparison for the Middle Pliocene. *Global Ecology and Biogeography*, *17*(3), 432–447. https://doi.org/10.1111/j.1466-8238.2008.00381.x
139. Sasser, C. E., Gosselink, J. G., Swenson, E. M., & Evers, D. E. (1995). Hydrologic, vegetation, and substrate characteristics of floating marshes in sediment-rich wetlands of the Mississippi river delta plain, Louisiana, USA. *Wetlands Ecology and Management*, *3*(3), 171–187. https://doi.org/10.1007/BF00180023
140. Sasser, C. E., Gosselink, J. G., Swenson, E. M., Swarzenski, C. M., & Leibowitz, N. C. (1996). Vegetation, substrate and hydrology in floating

marshes in the Mississippi river delta plain wetlands, USA. *Vegetatio*, *122*(2), 129–142. https://doi.org/10.1007/BF00044695

141. Sawyer, J. (1992). Timber Trade. In J. A. Sayer, C. S. Harcourt, & N. M. Collins (Eds.), *The Conservation Atlas of Tropical Forests Africa* (pp. 56–61). Palgrave Macmillan UK. https://doi.org/10.1007/978-1-349-12961-4_7
142. Scott, D. A., & Baer, S. G. (2018). Degraded soil increases the performance of a dominant grass, Andropogon gerardii (Big bluestem). *Plant Ecology*, *219*(8), 901–911. https://doi.org/10.1007/s11258-018-0844-0
143. Skagen, S. K., Burris, L. E., & Granfors, D. A. (2016). Sediment Accumulation in Prairie Wetlands under a Changing Climate: The Relative Roles of Landscape and Precipitation. *Wetlands*, *36*(2), 383–395. https://doi.org/10.1007/s13157-016-0748-5
144. Sloan, S., & Sayer, J. A. (2015). Forest Resources Assessment of 2015 shows positive global trends but forest loss and degradation persist in poor tropical countries. *Forest Ecology and Management*, *352*, 134–145. https://doi.org/10.1016/j.foreco.2015.06.013
145. Smardon, R. (2009). *Sustaining the World's Wetlands*. Springer US. https://doi.org/10.1007/978-0-387-49429-6
146. Soledad Ureta-Zañartu, M. (2022). How chemistry shaped societies: The case of the Atacama Desert in Chile. *ChemTexts*, *9*(1), 1. https://doi.org/10.1007/s40828-022-00172-3
147. Stoffel, M., Tiranti, D., & Huggel, C. (2014). Climate change impacts on mass movements—Case studies from the European Alps. *Science of The Total Environment*, *493*, 1255–1266. https://doi.org/10.1016/j.scitotenv.2014.02.102
148. Stott, R. S., & Olson, D. P. (1973). Food-Habitat Relationship of Sea Ducks on the New Hampshire Coastline. *Ecology*, *54*(5), 996–1007. https://doi.org/10.2307/1935566
149. Stuart, A. J. (1991). Mammalian Extinctions in the Late Pleistocene of Northern Eurasia and North America. *Biological Reviews*, *66*(4), 453–562. https://doi.org/10.1111/j.1469-185X.1991.tb01149.x
150. Syvitski, J. P. M., Overeem, I., Brakenridge, G. R., & Hannon, M. (2012). Floods, floodplains, delta plains—A satellite imaging approach. *Sedimentary Geology*, *267–268*, 1–14. https://doi.org/10.1016/j.sedgeo.2012.05.014
151. Tabata, S. (1975). The general circulation of the Pacific Ocean and a brief account of the oceanographic structure of the North Pacific Ocean Part I

- circulation and volume transports. *Atmosphere*, *13*(4), 133–168. https://doi.org/10.1080/00046973.1975.9648394

152. Tang, C. Q. (2015). Evergreen Broad-Leaved Forests. In C. Q. Tang (Ed.), *The Subtropical Vegetation of Southwestern China: Plant Distribution, Diversity and Ecology* (pp. 49–112). Springer Netherlands. https://doi.org/10.1007/978-94-017-9741-2_2
153. Tang, C.-S., Zhu, C., Cheng, Q., Zeng, H., Xu, J.-J., Tian, B.-G., & Shi, B. (2021). Desiccation cracking of soils: A review of investigation approaches, underlying mechanisms, and influencing factors. *Earth-Science Reviews*, *216*, 103586. https://doi.org/10.1016/j.earscirev.2021.103586
154. Tang, L., & Werner, T. T. (2023). Global mining footprint mapped from high-resolution satellite imagery. *Communications Earth & Environment*, *4*(1), Article 1. https://doi.org/10.1038/s43247-023-00805-6
155. Tchebakova, N. M., Parfenova, E. I., & Soja, A. J. (2016). Significant Siberian Vegetation Change is Inevitably Brought on by the Changing Climate. In L. Mueller, A. K. Sheudshen, & F. Eulenstein (Eds.), *Novel Methods for Monitoring and Managing Land and Water Resources in Siberia* (pp. 269–285). Springer International Publishing. https://doi.org/10.1007/978-3-319-24409-9_10
156. Thrush, S., Hewitt, J., Cummings, V., Ellis, J., Hatton, C., Lohrer, A., & Norkko, A. (2004). Muddy waters: Elevating sediment input to coastal and estuarine habitats. *Frontiers in Ecology and the Environment*, *2*(6), 299–306. https://doi.org/10.1890/1540-9295(2004)002[0299:MWESIT]2.0.CO;2
157. Tibbetts, J. R., & van Proosdij, D. (2013). Development of a relative coastal vulnerability index in a macro-tidal environment for climate change adaptation. *Journal of Coastal Conservation*, *17*(4), 775–797. https://doi.org/10.1007/s11852-013-0277-9
158. Toorman, E. A., Anthony, E., Augustinus, P. G. E. F., Gardel, A., Gratiot, N., Homenauth, O., Huybrechts, N., Monbaliu, J., Moseley, K., & Naipal, S. (2018). Interaction of Mangroves, Coastal Hydrodynamics, and Morphodynamics Along the Coastal Fringes of the Guianas. In C. Makowski & C. W. Finkl (Eds.), *Threats to Mangrove Forests: Hazards, Vulnerability, and Management* (pp. 429–473). Springer International Publishing. https://doi.org/10.1007/978-3-319-73016-5_20
159. Valencia-Aguilar, A., Cortés-Gómez, A. M., & Ruiz-Agudelo, C. A. (2013). Ecosystem services provided by amphibians and reptiles in Neotropical ecosystems. *International Journal of Biodiversity Science,*

Ecosystem Services & Management, *9*(3), 257–272. https://doi.org/10.1080/21513732.2013.821168

160. van der Veen, C. J. (1999). Crevasses on glaciers 1. *Polar Geography*, *23*(3), 213–245. https://doi.org/10.1080/10889379909377677
161. Vernon, M. E., Campos, B. R., & Burnett, R. D. (2022). Effects of Livestock Grazing On The Ecology Of Sierra Meadows: A Review of The Current State of Scientific Knowledge To Inform Meadow Restoration And Management. *Environmental Management*, *69*(6), 1118–1136. https://doi.org/10.1007/s00267-022-01634-7
162. Viereck, L. A., Van Cleve, K., & Dyrness, C. T. (1986). Forest Ecosystem Distribution in the Taiga Environment. In K. Van Cleve, F. S. Chapin, P. W. Flanagan, L. A. Viereck, & C. T. Dyrness (Eds.), *Forest Ecosystems in the Alaskan Taiga: A Synthesis of Structure and Function* (pp. 22–43). Springer. https://doi.org/10.1007/978-1-4612-4902-3_3
163. Vinson, M. R., & Hawkins, C. P. (2003). Broad-scale geographical patterns in local stream insect genera richness. *Ecography*, *26*(6), 751–767. https://doi.org/10.1111/j.0906-7590.2003.03397.x
164. Voigt, P. W., Rethman, N. F. G., & Poverene, M. M. (2004). Lovegrasses. In *Warm-Season (C4) Grasses* (pp. 1027–1056). John Wiley & Sons, Ltd. https://doi.org/10.2134/agronmonogr45.c32
165. Wang, G., Hu, H., & Li, T. (2009). The influence of freeze–thaw cycles of active soil layer on surface runoff in a permafrost watershed. *Journal of Hydrology*, *375*(3), 438–449. https://doi.org/10.1016/j.jhydrol.2009.06.046
166. Wantzen, K. M., Marchese, M. R., Marques, M. I., & Battirola, L. D. (2016). Invertebrates in Neotropical Floodplains. In D. Batzer & D. Boix (Eds.), *Invertebrates in Freshwater Wetlands: An International Perspective on their Ecology* (pp. 493–524). Springer International Publishing. https://doi.org/10.1007/978-3-319-24978-0_14
167. Weller, G. (2000). The Weather and Climate of the Arctic. In *The Arctic*. Routledge.
168. Whittaker, R. J. (1993). Plant population patterns in a glacier foreland succession: Pioneer herbs and later-colonizing shrubs. *Ecography*, *16*(2), 117–136. https://doi.org/10.1111/j.1600-0587.1993.tb00064.x
169. Xin, P., Wilson, A., Shen, C., Ge, Z., Moffett, K. B., Santos, I. R., Chen, X., Xu, X., Yau, Y. Y. Y., Moore, W., Li, L., & Barry, D. A. (2022). Surface Water and Groundwater Interactions in Salt Marshes and Their Impact on Plant Ecology and Coastal Biogeochemistry. *Reviews of Geophysics*, *60*(1), e2021RG000740. https://doi.org/10.1029/2021RG000740

170. Young, N. B. M., Kenneth R. (2020). Savannas and Grasslands. In *Terrestrial Ecosystems and Biodiversity* (2nd ed.). CRC Press.
171. Zech, W., Guggenberger, G., Haumaier, L., Pöhhacker, R., Schäfer, D., Amelung, W., Miltner, A., Kaiser, K., & Ziegler, F. (1996). Chapter 3—Organic Matter Dynamics in Forest Soils of Temperate and Tropical Ecosystems. In A. Piccolo (Ed.), *Humic Substances in Terrestrial Ecosystems* (pp. 101–170). Elsevier Science B.V. https://doi.org/10.1016/B978-044481516-3/50004-9
172. Zech, W., Schad, P., & Hintermaier-Erhard, G. (2022). Dry Mid-Latitudes. In W. Zech, P. Schad, & G. Hintermaier-Erhard (Eds.), *Soils of the World* (pp. 77–98). Springer. https://doi.org/10.1007/978-3-540-30461-6_5
173. Zeng, H., & Jia, G. (2013). Impacts of snow cover on vegetation phenology in the arctic from satellite data. *Advances in Atmospheric Sciences*, *30*(5), 1421–1432. https://doi.org/10.1007/s00376-012-2173-x
174. Zhang, X., Zhou, Y., Ji, Y., Yu, M., Li, X., Duan, J., Wang, Y., Gao, J., & Guo, X. (2023). Climate Factors Affect Above–Belowground Biomass Allocation in Broad-Leaved and Coniferous Forests by Regulating Soil Nutrients. *Plants*, *12*(23), Article 23. https://doi.org/10.3390/plants12233926
175. Zubairov, B., Lentschke, J., & Schröder, H. (2019). Dendroclimatology in Kazakhstan. *Dendrochronologia*, *56*, 125602. https://doi.org/10.1016/j.dendro.2019.05.006

INDEX

O

P

R

S

T

U

V

W

Z

FSC
www.fsc.org
MIX
Papier | Fördert gute Waldnutzung
FSC® C083411

Kolibri 360 GmbH
Ferdinand-Jühlke-Straße 7
99095 Erfurt, Deutschland
produktsicherheit@kolibri360.de